Furan Polymers and their Reactions

Furan Polymers and their Reactions

Alessandro Gandini

Retired from the Polytechnic Institute
Grenoble
France

Talita Martins Lacerda

University of São Paulo
Lorena
Brazil

WILEY

This edition first published 2024

Registered Office
John Wiley & Sons Ltd, The Atrium, Southern Gate, Chichester, West Sussex, PO19 8SQ, UK

For details of our global editorial offices, customer services, and more information about Wiley products visit us at www.wiley.com.

Wiley also publishes its books in a variety of electronic formats and by print-on-demand. Some content that appears in standard print versions of this book may not be available in other formats.

A catalogue record for this book is available from the Library of Congress

Hardback ISBN: 9781119888697; ePub ISBN: 9781119888703; ePDF ISBN: 9781119888680; oBook ISBN: 9781119888710

Cover Image: Courtesy of Talita Martins Lacerda
Cover design by Wiley

Set in 9.5/12.5pt STIXTwoText by Integra Software Services Pvt. Ltd, Pondicherry, India
Printed and bound by CPI Group (UK) Ltd, Croydon, CR0 4YY

C9781119888697_231123

Contents

Foreword

Furan Polymers and their Reactions surveys the chemistry of these crucial renewable heterocyclic compounds and their contributions to polymer synthesis with a large variety of their structures. This book discusses the biorefinery of furans, identifies furfural and 5-hydroxymethyl furfural as the key furan monomer precursors for different polymer synthetic processes, and analyzes all the major reactions the monomers undergo during these processes, as well as the structures, properties, modifications, and applications of the ensuing materials. The results are a vital contribution to the growing field of renewable macromolecular materials.

The book provides an up-to-date and detailed analysis of key polymerization reactions such as chain-growth and step-growth polymerizations and the furan/maleimide Diels–Alder reaction, both as a polycondensation tool and as one of the sources of the chemical modifications of polymers.

Furan Polymers and their Reactions is an essential resource and is highly recommended for researchers and professionals in industrial engineering, polymer science, and biotechnology, as well as for any industry professionals working with platform chemicals or polymer synthesis.

Kris Matyjaszewski
Carnegie Mellon University
Pittsburgh, May 2023

Preface

Eons of evolution have gradually transformed biological structures and functions in all areas of life through increasing molecular complexity and interactions which provided specific adaptations by natural selection. The synthesis of macromolecules in both animal and vegetal species has fulfilled vastly different roles ranging from DNA for genetic information to cellulose and lignin for structural solidity, all being regularly renewed by appropriately adapted pathways in which solar energy is, directly or indirectly, the driving force. In the context of materials science, natural polymers have provided mankind for millennia with valuable sources of living support, including shelter, clothing, tools, and cultural artefacts, aided by technological adaptations, such as natural fiber weaving, papyrus- and then papermaking, leather tanning and construction ingenuities like waterproofing, gluing, as well as hunting and fishing aids. Although some chemistry was unknowingly involved in these activities, the actual suitable chemical modification of natural polymers only begun in the nineteenth century essentially with the preparation of cellulose esters, the vulcanization of rubber, the exploitation of natural resins for making adhesives and sealants, and the controlled drying of plant oils to prepare plastic sheets like linoleum. The major progress, however, occurred throughout the twentieth century because of the chemical revolution associated with the exploitation of fossil resources to prepare monomers and to implement their polymerizations and thus provide an array of synthetic macromolecular materials endowed of different properties and applications. This vigorous development, particularly during the second half of the century, which took the inaccurate name of "plastic revolution", went for a large part to the detriment of the pursuit of research in the realm of renewable resources as providers of monomers and their polymers, with few novel contributions. This situation started changing toward the end of the century when the paradigm of sustainability begun to make its way within the general conscience in the wake of calls to reduce the dependence on materials and energy providers based on the use of dwindling petroleum and natural gas reserves and to reduce their negative ecological impact to the planet. Research on the alternative exploitation of renewable resources for the same purposes thus resurfaced first timidly, then progressively with more impetus in academy and industry as the century was coming to a close to welcome the new millennium.

This state of affairs is opening a bright scenario in which polymers from renewable resources are witnessing a spectacular growth in the twenty-first century because they represent the response of macromolecular science and technology to the challenges of fighting climate deregulation, together with the equally fundamental search for renewable energy resources, both presently dominated by the exploitation of fossil counterparts. A 2008 monograph provided the first comprehensive state of the art on these materials [1] and numerous reviews followed, with a recent overview of the topic [2] aimed at highlighting the incessant progresses and perspectives in all its domains. The topic is usually subdivided into three major approaches to the elaboration of materials from renewable resources, namely (i) the synthesis of important polymers which are traditionally prepared from fossil resources and can also be prepared now using the same monomers, but derived from renewable counterparts, (ii) the chemical modification of natural polymers to generate novel materials, and (iii) the exploitation of a variety of monomers and macromonomers derived from renewable resources to prepare original macromolecular structures with interesting, or unique, properties and promising applications.

Furans constitute a major class of potential monomers pertaining to the third approach, and this book's title is a minimalist condensation of the important reality of polymers containing furan units derived from natural resources based on polysaccharides and sugars found worldwide as agricultural and forestry side-products, or available in sufficient quantities to justify their exploitation without affecting the food and feed requirements. Two compounds, furfural and hydroxymethylfurfural are readily prepared from these pentoses and hexoses and can be turned into a variety of monomers, thus generating a whole family of sustainable materials covering virtually all structures and properties already found in fossil-based polymers and even previously unattained performances. This situation is unique, since it cannot be achieved with other natural polymer like cellulose, starch, natural rubber and chitosan, or indeed with such natural monomers as terpenes and vegetable oils, because in all these contexts, their inherent chemical structures limits the breadth of polymer properties that can be obtained, notwithstanding the very impressive variety of materials which have and are being successfully prepared and which have so importantly enriched the domain of polymers from renewable resources.

Molecules and macromolecules incorporating the furan heterocycle have been known or synthesized since 1780 when 2-furoic acid was first prepared by Scheele, followed by furfural, furfuryl alcohol and furan itself in the nineteenth century, and by hydroxymethylfurfural and the first industrial furfuryl alcohol resins in the mid-twentieth century. The real event, however, that established furans as a widely recognized branch of chemistry was the publication in 1953 of a seminal book, *The Furans*, by A.P. Dunlop and F.N. Peters (ACS Monograph Series N° 119, Reinhold Publishing Co., New York), two high ranking industrial chemists working at the Quaker Oats Company, the progenitor of furfural and furfuryl alcohol. This monumental monograph of close to 900 pages and covering virtually all furan compounds known then, set the scenario for the development of furans in all major branches of chemistry, including organic synthesis, natural compounds, pharmaceuticals, industrial aspects (particularly for the production of furfural), and the nascent macromolecular field (particularly related to furfuryl alcohol resins). Although numerous book chapters, reviews and patents, based on much published research, have since been covering these topics, no other volume has appeared updating the general field of furans in such a comprehensive treatment. Indeed, the present book is no exception, because its purpose is confined to the area of furan polymers, with the specific message of highlighting the renewable character of these materials and hence providing a contribution to the enhancement of sustainability in the realm of macromolecular science. The field of furan polymers is less than a century old and much younger if one considers materials other than the by-now classic resins based on furfuryl alcohol. It seems fair to point out that serious interest in these materials only begun a few decades ago. Several reviews have covered this area since then but, given the rapid upsurge of important advances in different widening contexts, the need for a more comprehensive and updated monograph prompted us to write this book, which attempts to provide a systematic coverage, while giving more emphasis to recent progress, and to propose ideas about future perspectives and practical developments.

We thought it would be useful to begin with a brief historical sketch and with a chapter devoted to a concise reminder of the chemical peculiarities associated with the furan heterocycle because of their strong implications in determining the behavior and mechanisms of most of the types of polymerizations to be discussed. This introductory chapter is however not intended to cover the organic chemistry of furans, whose synthetic and reactivity aspects fall outside the scope of the book. We also decided to treat the polycondensation of furfuryl alcohol and the ensuing materials in a separate first chapter of the book's main subject because of the unique historical and mechanistic features of what remains today the most important furan industrial material, although this primacy might not last much longer. The rest of the book is organized as in a classical polymer treatise, namely by treating chain- and step-growth systems separately. The properties and potential applications of these materials from renewable resources are outlined when the preparation of each of them is discussed, often comparing their features with those of traditional fossil-based counterparts or emphasizing their unique connotations. We wish to reiterate that the sustainability connotation permeating each section of the book is as essential an ingredient as its actual portrayal of a notable family of macromolecular materials.

References

1 Belgacem, M.N. and Gandini, A. (Eds.) (2008). *Monomers, polymers and composites from renewable resource*. Elsevier, Amsterdam.

2 Gandini, A. and Lacerda, T.M. (2022). Polymers from renewable resources: Macromolecular materials for the twenty-first century? In: *Macromolecular Engineering: From Precise Synthesis to Macroscopic Materials and Applications*, 2e (eds. K. Matyjaszewski, Y. Gnanou, N. Hadjichristidis, and M. Muthukumar). WILEY-VCH GmbH, Vol. 5, 1–77.

1
A Brief History

Furfural (F) industrial production, from pentose-rich oat hulls, begun in 1922 at the Quaker Oats cereal mill in Iowa, and soon after its first resins for molding and abrasive tools were on the market in the US. That was followed by furfuryl alcohol (FA) industrial production by the same company in 1934, through an efficient F reduction process, and its resins for the foundry business in 1958 became commercially available. In both instances, these materials were crosslinked polymers with useful thermal and mechanical properties, but little was known about their mechanisms of formation and ultimate structures. It is most likely that the resinification of both these furans was a frequent unwanted event when handling them from their earlier synthetic operations and isolation, given their sensitivity to accidental polymerization, particularly in acidic media. It is moreover particularly relevant to note that since the inception of a series of thermoplastic (cellulose esters) and thermoset (linoleum and vulcanized natural rubber) materials from renewable resources during the second half of the nineteenth century, these furan resins were the first novel materials being produced from renewable resources in the twentieth century. In another vein, F was converted into adiponitrile for the manufacture of Nylon 6,6 and in this du Pont process of 1949, the tetrahydrofuran intermediate was also later utilized to prepare poly(tetramethylene ether)glycol.

A very modest number of other interesting contributions to furan polymers can be found in the period going back to the late 1960s, whereas the following decade saw what should be considered as the beginning of serious research aimed at preparing and characterizing such materials and unravelling the mechanisms of their formation. The first review on furan polymers, published by one of us in 1977 [1], reflects this initial ferment, albeit underlining that much of it was not entirely gratifying in terms of scientific depth or practical success as far the properties of the ensuing materials. It can be argued that this state of affairs stemmed in part from the few and non-communicating groups scattered around the world, with only one laboratory at Havana University doing systematic work on the

Furan Polymers and their Reactions, First Edition. Alessandro Gandini and Talita Martins Lacerda.

polymerization of alkenyl furans and other F-derived monomers. Examples of the paucity of research topics pointed out in that review reflected mostly the rare reports on furan polyesters and polyamides and the absence of studies on polyurethanes, epoxies and other relevant polycondensates. This problem mostly reflected the difficulty in efficiently preparing bifunctional monomers, which were difficult to attain from the monofunctional F and FA.

Some 10 years later, a second review [2] reflected a much-improved situation with a variety of novel investigations covering a wider domain of polymer syntheses and structures, suggesting that the 1980s had brought about a significant awakening concerning the interest of working on furan polymers. But the availability of bifunctional monomers persisted because of the difficulties associated with producing hydroxymethylfurfural (HMF) from hexoses in economically viable quantities.

The overall positive trend was however to continue, as reflected by a third review published a decade later [3], in which the sheer volume of available scientific literature on the subject (reflected by the more than 300 references) was as impressive as much of its scientific and applied quality. This advance was facilitated by the progress in the exploitation of HMF as a source of various bifunctional monomers well suited for the synthesis of polyesters, polyamides and polyurethanes, among other polycondensates.

The research aimed at optimizing the production of HMF became an international effort at the end of the twentieth century, with the primary aim consisting in using it to prepare the corresponding diacid, 2,5-furandicarboxylic acid (FDCA) as the priority monomer, although other difunctional homologues, such as the 2,5-bis(hydroxymethyl)furan (BHMF), were also actively sought, as discussed in a fourth review [4]. The success of this strategy and of other original approaches, notably the application of the Diels–Alder reaction, meant that within another quarter of a century, furan polymers have progressively become a household item to macromolecular scientists, as highlighted by a recent fifth review [5], which represented a skeletal work in progress for the writing of this comprehensive book, which attempts to provide a full picture, going from the pioneering investigations of some 50 years ago to the opening up of brilliant perspectives inspired by the most recent advances.

The numerous breakthroughs that took place in this relatively short saga are documented in detail in each chapter together with a deserved mention to the colleagues who were and are behind these achievements.

References

1 Gandini, A. (1977). *Adv. Polym. Sci.* 25: 47–96.
2 Gandini, A. (1986). *Encyclopedia Polym. Sci. Eng.* 7: 454–473.
3 Gandini, A. and Belgacem, M.N. (1997). *Progr. Polym. Sci.* 22: 1203–1379.
4 Moreau, C., Gandini, A. and Belgacem, M.N. (2004). *Topics Catal.* 27: 11–30.
5 Gandini, A. and Lacerda, T.M. (2022). *Macromol. Mater. Eng.* 307: 2100902.

2

The Peculiar Chemical Features of the Furan Heterocycle and the Synthesis of Furfural and Hydroxymethylfurfural

Following the classic treatise where Dunlop and Peters covered systematically in 1953 the domain of furans in a comprehensive approach [1], several monographs discuss advances in the organic chemistry of furans [2], their naturally occurring compounds [3], and the pharmacological features of some derivatives [4]. This book does not deal with these issues, concentrating instead on furan polymers, but a brief outlook is provided here about the furan-specific chemical properties and reactivity, which are closely related to the polymerization behavior of furan monomers.

The furan heterocycle belongs to the family of five-member ring unsaturated compounds, and its structure can be displayed by the resonant forms shown in Figure 2.1.

Substitution reactions of furans such as alkylation, halogenation, nitration, and sulfonation take place regioselectively at the C2 and/or C5 positions, which indicate that C and D are the dominant structures.

Compared with its sulfur and nitrogen homologues, furan possesses a reduced aromatic character and thus the most dienic, as illustrated in Figure 2.2, where benzene is the reference aromatic compound and cyclopentadiene the reference diene. The Dewar resonance energies decrease from 95 kJ mol^{-1} for benzene to 27 for thiophene, 22 for pyrrole, 18 for furan, and close to zero for cyclopentadiene.

This feature has an important bearing on much of the chemical behavior of furan, as briefly outlined below.

2.1 Free Radical Reactions

The addition of a free radical to a C2-substituted furan heterocycle takes place predominantly at the C2 or C5 position, as shown in Figure 2.3, depending on the chemical nature and steric hindrance of the substituent. The ensuing furyl radical unpaired electron is shared by the remaining three carbon atoms of the ring

Furan Polymers and their Reactions, First Edition. Alessandro Gandini and Talita Martins Lacerda.

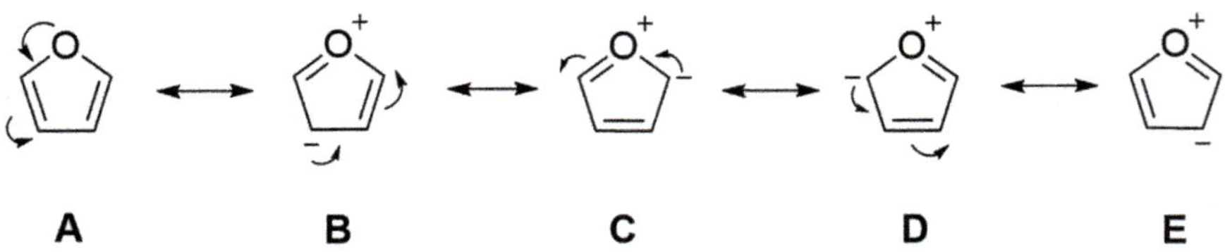

Figure 2.1 The resonance-contributing structures associated with the furan heterocycle.

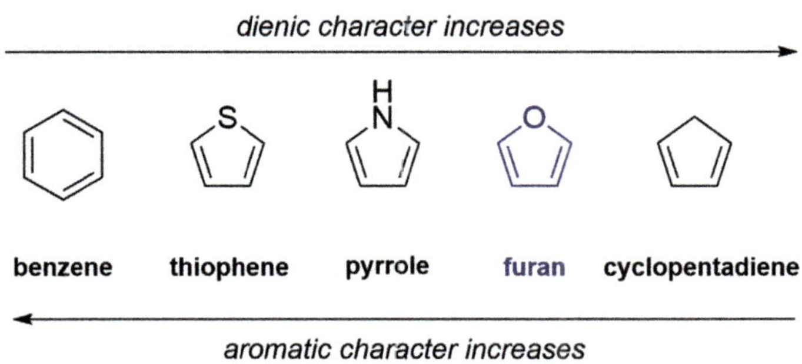

Figure 2.2 The relative dienic and aromatic character of thiophene, pyrrole, and furan in relation to the fully aromatic benzene and the fully dienic cyclopentadiene.

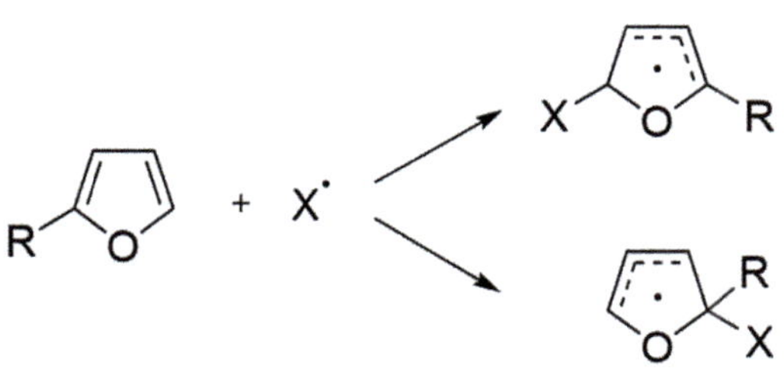

Figure 2.3 The addition modes of a primary radical to furan and 2-monosubstituted furan compounds.

according to its dienic character, which is expected to confer a certain stability to the structure. The reactions of these radicals derived from furan monomers will be discussed in detail when dealing with their behavior in polymerization systems, but their sluggish reactivity must already be mentioned here in the sense that they do not display any tendency to oligomerize with the excess precursor, which emphasize their considerable degree of stabilization. Their termination in a context where the furan derivative is only confronted with primary radicals are illustrated in Figure 2.4, with the formation of a 2,5-dihydrofuran, or the corresponding C5-C5 dimer, depending on the terminating radical. These results were obtained in a systematic study of the bulk reaction of AIBN with furan and a wide selection of furan 2-substituted derivatives [5].

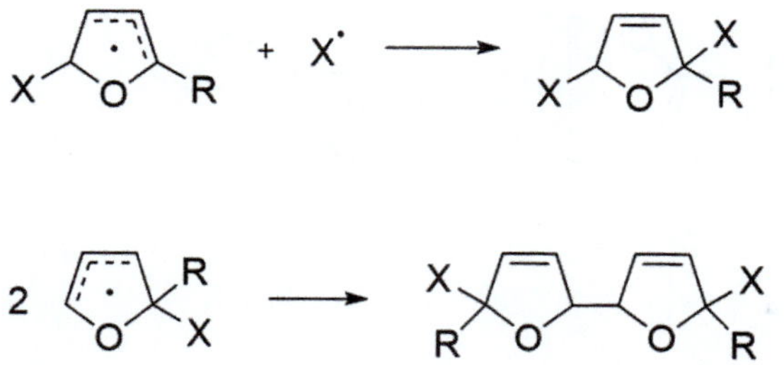

Figure 2.4 Two alternative termination reactions of 2-furyl radicals.

2.2 Electrophilic Reactions

The addition of an electrophile to furan and 2-monosubstituted furans occurs predominantly at the C5 position, as shown in Figure 2.5. This very pronounced regiospecificity, estimated to more than 100:1 with respect to reactions at C3 or C4, is attributed again to the dienic character of the heterocycle, and bears strong consequences to the behavior of the cationic polymerization of furan monomers discussed in the appropriate section. Suffice to note here that the carbocation intermediate in these reactions is stabilized by the dienic property of the furan ring.

2.3 Photochemistry

The electronic spectra of furan and alkyl furans display maxima at 205–215 nm associated with the conjugated diene $\pi \rightarrow \pi^*$ transition. Their mercury-sensitized photolysis produces both molecular fragmentation into CO, cyclopropenes, alkynes and allenes after ring contraction, and isomerization from the excited intermediates, as shown in Figure 2.6 for the case of 2-mehyl furan [6].

When a carbonyl function is attached to the heterocycle, the ring diene is now conjugated to it giving rise, in the case of furfural, to an electronic spectrum with a strong $\pi \rightarrow \pi^*$ transition peak at 260 (gas)-280 (solution) nm and a much weaker $\mathbf{n} \rightarrow \pi^*$ transition at 330 nm in the gas phase, which merges into the red-end of the $\pi \rightarrow \pi^*$ transition in the spectrum of liquid furfural. Furoic acid, HMF and 2-methylfuryl ketone exhibit similar spectral features, which have been exploited for the quantitative analysis of these compounds in processes involving their formation or presence as impurities.

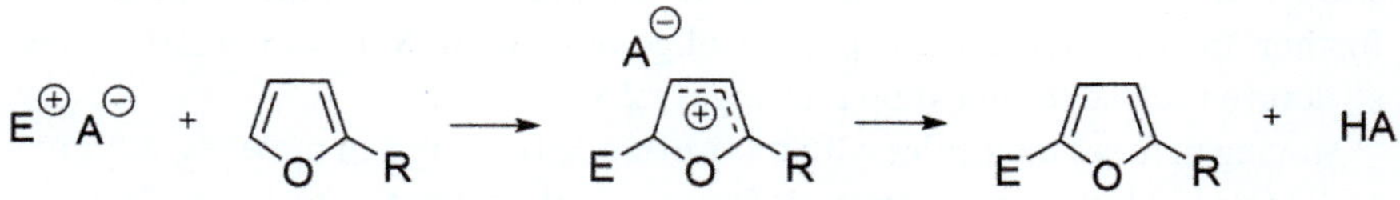

Figure 2.5 Electrophilic substitution reaction to a 2-substituted furan derivative.

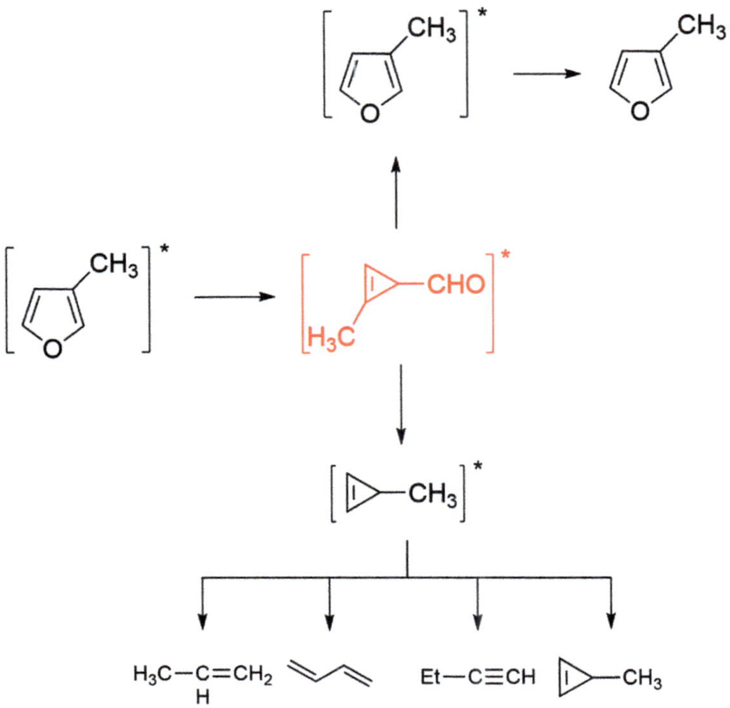

Figure 2.6 The mercury-sensitized photolysis of 2-methyl furan.

The photolysis of furfural in the gas phase has been studied both by direct irradiation of both bands and by mercury sensitization [7]. In all instances, the singlet and triplet excited states rapidly revert to a ground-state vibrationally excited molecule, which either loses a CO molecule to yield furan, or suffers ring contraction, as in the case of furan, to give a cyclopropene dialdehyde intermediate that decomposes into two molecules of CO, propyne, allene, and cyclopropene. At higher pressures, some oligomerization is also observed. On the whole, therefore, this photochemical behavior is similar to that of furan, illustrated in Figure 2.6.

On the other hand, when liquid furfural sealed within a quartz tube is exposed to the full spectrum of a medium-pressure mercury arc for 10–90 h [8], the major product is an oligomeric material generated by two alternative reactions involving furfural excited state and a ground-state molecule, as shown in Figure 2.7. The ensuing vibrationally excited dimer reacts with further furfural molecules to give oligomers with Mn = 500±50, whose structure is close to that shown in Figure 2.8.

Moving to furan molecules with more extended conjugated moieties, 2-furanacrylic acid has been reported to cyclodimerize in the solid state to give cyclobutane

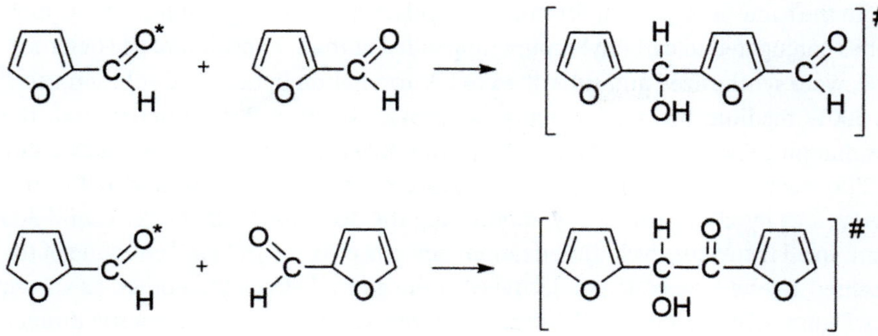

Figure 2.7 The interactions of electronically excited furfural molecules with ground-state counterparts.

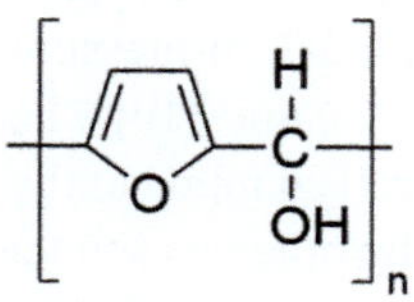

Figure 2.8 The structure of the repeat unit of furfural oligomers produced by liquid-phase photolysis.

isomers [9]. A water slurry of crystals of the corresponding 2,5-diacid (FDA) was irradiated with a 100 W tungsten bulb in a water-cooled reactor giving a near-quantitative conversion into the cyclodimer arising from the [π2+π2] coupling of the excited alkene moiety of a diacid molecule with the ground-state of the second one [10]. The structure of this product, shown in Figure 2.9, was confirmed by a thorough spectroscopic characterization. The use of a simple light bulb (and indeed also sunlight) was sufficient to promote this solid-state reaction stems from the fact that FDA displays a strong peak at 350 nm tailing off towards the visible domain, and photolysis with mercury lamps induced its oligomerization.

It is important to note that the dimer reverted to FDA when heated at 220 °C, a typical behavior related to the breaking down of the strained cyclobutane ring, which will be evoked in other similar systems.

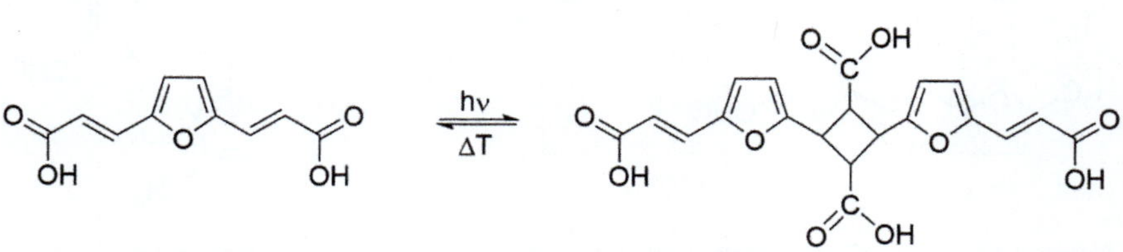

Figure 2.9 The photochemical dimerization of 2,5-furandiacrylic acid.

In the same vein, several furan compounds bearing conjugated *trans* C=C moieties conjugated to carbonyl groups, appended at the C2 position of the heterocycle, were synthesized and submitted to UV irradiation in bulk and solution using a 500W medium-pressure mercury arc provided with a Pyrex filter to limit the excitation to wavelengths higher than 270 nm [11a]. Reactions were followed by FTIR spectroscopy, monitoring the decrease in the C=C band around 1640 cm^{-1}, as well as by UV spectroscopy, monitoring the decrease of the peak around 300 nm. In all instances, cyclodimerization took place via the [π2+π2] coupling of the excited alkene moiety of a molecule with the ground-state of the other, as shown in Figure 2.10, which also illustrates the formation of the two isomeric dimers, identified spectroscopically.

In a systematic study on the photodimerization of furylene-vinylene and thienylene-vinylene compounds, the dimer arising from the self-condensation of 5-methyl furfural (5MF), which gave a UV maximum at 393 nm in methylene chloride, was irradiated in the solid state with both high- and medium-pressure mercury arcs using a filter to cut out any wavelength lower than 300 nm [11b]. The compound readily yielded the symmetrical dimer via the [π2+π2] coupling of the excited alkene moiety of a molecule with the ground-state of the other, as shown in Figure 2.11, which was fully characterized. This reaction was accompanied by

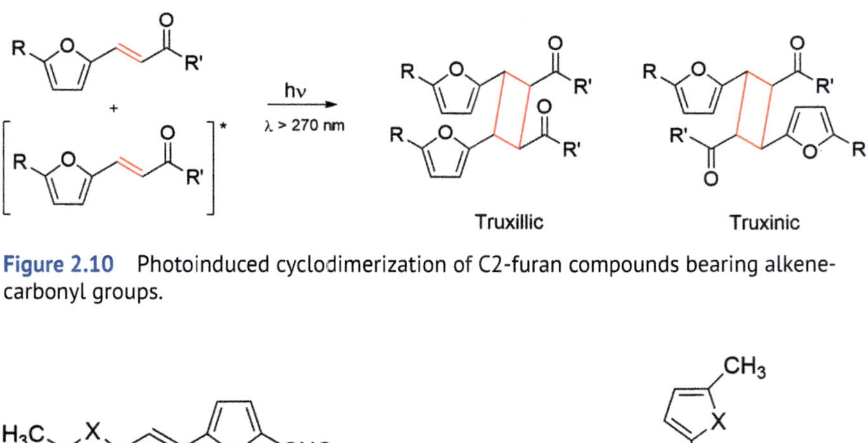

Figure 2.10 Photoinduced cyclodimerization of C2-furan compounds bearing alkene-carbonyl groups.

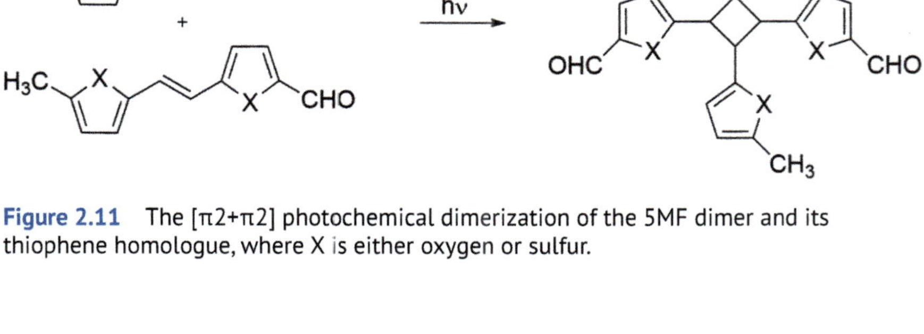

Figure 2.11 The [π2+π2] photochemical dimerization of the 5MF dimer and its thiophene homologue, where X is either oxygen or sulfur.

some trans-cis isomerization of the irradiated molecules. Again, it was observed that heating the crystalline dimer produced a reversion to its precursor starting at about 170 °C. The other furan and thiophene structures bearing ester, and pyridine moieties behaved likewise, but gave rise to a mixture of truxinic and truxillic isomeric dimers.

All these investigations constitute an important basis for their extension to polymerization and polymer cross-linking systems calling upon photochemical activation, as discussed in the specific chapters later in this book.

2.4 Hydrolysis Reactions

The furan heterocycle is susceptible to acid-catalyzed hydrolysis, requiring however fairly aggressive conditions in terms of a large excess of water and a very low pH. The reactions characterizing these processes are illustrated schematically in Figure 2.12. The synthetic relevance of hydrolysis has been exploited with the aim of preparing aliphatic carbonyl compounds. On the whole, however, the vast majority of the polymerization systems discussed in the book do not take place in such environments and are therefore virtually exempt from the interference of ring hydrolysis as an unwanted side reaction. Only when such drastic conditions are involved, the problem will therefore be highlighted, as in the case of polycondensation reactions catalyzed by strong acids that release water molecules at each growth step.

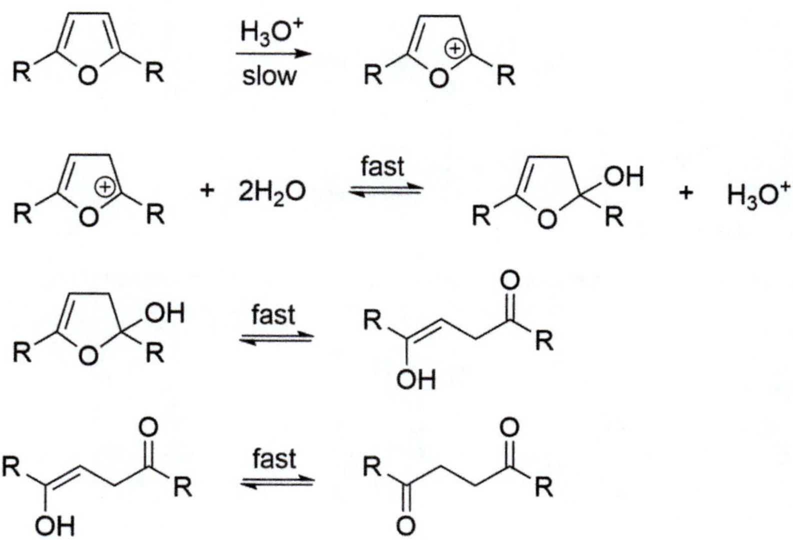

Figure 2.12 The typical reactions associated with the acid-catalyzed hydrolysis of furans.

2.5 The Diels–Alder Reaction

In the late 1920s and early 1930s, Otto Diels and Kurt Alder carried out a historical series of studies involving the coupling of dienes with dienophiles, which gave rise to the classical generic mechanism which took their names, usually abbreviated as the DA reaction. The DA reaction is a [4 + 2] cycloaddition between an electron-rich diene and an electron-poor dienophile. This reaction was later categorized under the family of "click" reactions since it embodies several attributes such as high yield, high specificity, lack of by-product and side reactions, and no essential need of a catalyst. Interestingly, the 2022 Nobel prize in Chemistry was awarded to three brilliant pioneers of click chemistry research. The DA reaction is one of the most powerful and versatile reactions in the toolbox of the synthetic organic chemist, allowing complex architectures to be constructed from simple building blocks [12].

A classical illustration of the DA reaction mechanism is given in Figure 2.13 for the coupling of two alkenes possessing complementary electronic properties.

The furan heterocycle, with its characteristic dienic properties, has been extensively exploited in the general context of DA chemistry [12] and the much-studied reaction of furan with maleic anhydride, displayed in Figure 2.14, illustrates its features, including the formation of the endo- and exo-isomeric adducts.

The DA reaction of furans with maleimides, shown in Figure 2.15, represents the starting point of a vigorous development of a new branch of furan polymer science which will be amply discussed later in the book. A systematic study of this reaction involving a large variety of furan and maleimide derivatives [13]

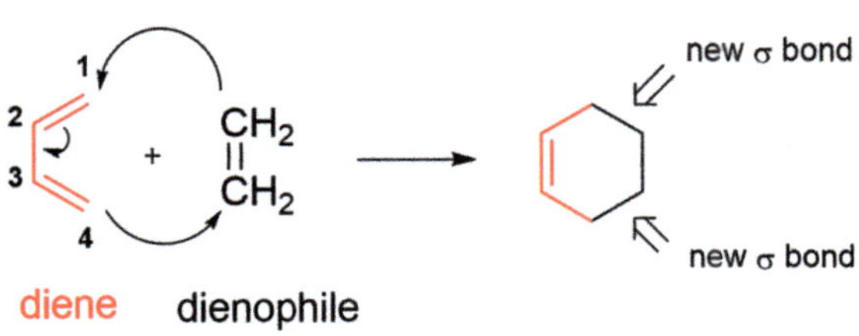

Figure 2.13 The Diels–Alder reaction mechanism between butadiene and ethylene.

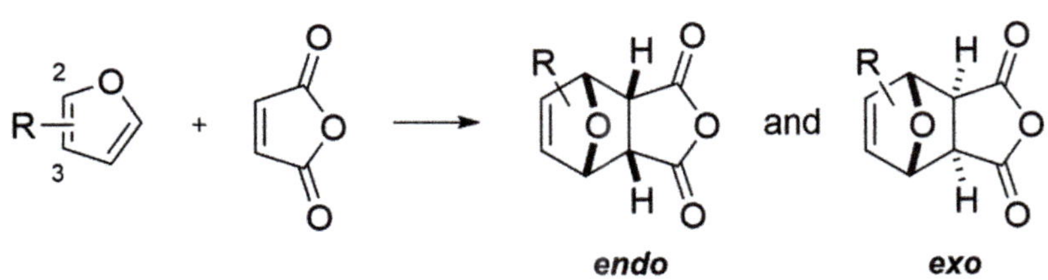

Figure 2.14 The DA reaction of furan monosubstituted compounds with maleic anhydride.

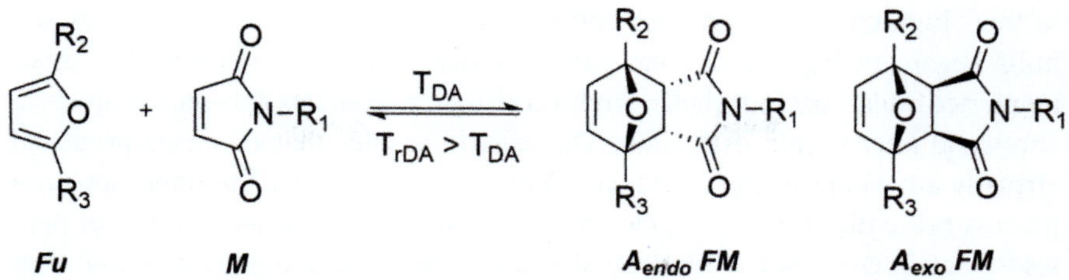

Figure 2.15 The DA reaction of furan derivatives (Fu) with maleimides (M) to give thermally reversible adducts (A).

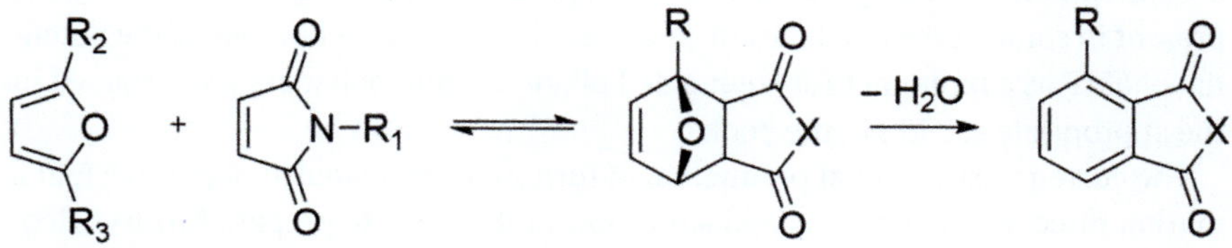

Figure 2.16 The aromatization of DA adducts (X = O or NH).

examined its kinetic and thermodynamic features and emphasized its thermal reversibility at experimentally accessible temperatures. This investigation confirmed the notion that carbonyl-substituted furans, like furfural and furoic acid, are not sufficiently reactive because of their poor diene connotation. This notion has however been challenged by recent studies in which ingenious experimental adaptations demonstrated that furan carbonyl derivatives could be made to react successfully [14]. The vast implication of this reversal of reactivity will be emphasized when discussing the application of the DA reaction to furan polymers.

The thermal reversibility of these adducts constitutes a key feature which is rather unique in the DA chemistry and has been largely exploited in polymer science, as discussed in the appropriate section of this book, together with the fact that the use of renewable furan reagents is synonymous of a fundamental contribution to sustainability. This reversibility can however be circumscribed if the adducts are deemed useful precursors to other structures. In particular, a simple dehydration reaction can be exploited to convert them irreversibly to aromatic compounds, as shown in Figure 2.16.

2.6 Furfural and Hydroxymethylfurfural as Industrial Commodities and as Building Blocks for Furan Monomers

One of the remarkable aspects of furfural (F) potential as a powerful renewable resource, and not only for its role of polymer building block, is the large number of agricultural and forestry side products which can be used to produce it thanks to their high content of pentose polysaccharides. These include corn cobs, oat hulls, sugarcane bagasse, almond husks, cottonseed hulls, residues of olive treatment, rice hulls, hazelnut shells, birchwood, beechwood, flax shivers, and quebracho wood after tannin extraction. This clearly implies that F can be produced virtually anywhere in the world, considering moreover that the only industrial process presently in use, the acid-catalyzed hydrolysis and dehydration of pentoses (mainly D-xylose), is relatively simple and hence accessible even to countries with a modest technological development [1,15]. Figure 2.17 gives a schematic sequence of the three-step dehydration process leading to F from the C5 sugar. 5MF is sometimes a by-product of furfural production, arising from L-rhamnose, present in some of the cited natural sources. This is relevant because of the potential of 5MF as a monomer for conjugated oligomers and polymers, as discussed in the appropriate section in the book.

The current yearly global production of furfural is estimated at over 440 kT, at a market price of about 1 $/kg, and some 85% of it is used to prepare furfuryl alcohol. It is important to emphasize that up to a few years ago, furfural was the only industrial furan derivative and hence the precursor to all synthetic furans.

Hydroxymethylfurfural (HMF) is the new star in the furan polymer firmament because it has been opening up new avenues for the synthesis of a blooming family of promising materials. Although it was a well-known compound for decades, recent research efforts are producing a growing success in achieving its economically viable industrial production, which was reported in several instances in different countries from 2013 [16]. HMF can be synthesized from a variety of C6 polysaccharides, like starch, or sugars, like fructose, as sketched by the sequence of steps given in Figure 2.18. This triple dehydration mechanism is entirely similar to that leading to F, but the major difference which made it cumbersome to implement it industrially stems from the chemical fragility of HMF, which is prone to suffer structural modification and/or resinification that affect its yield and purification. The possibility of oxidizing it to 2,5-furandicarboxylic acid (FDCA) in a one-pot process during its synthesis seems a particularly interesting

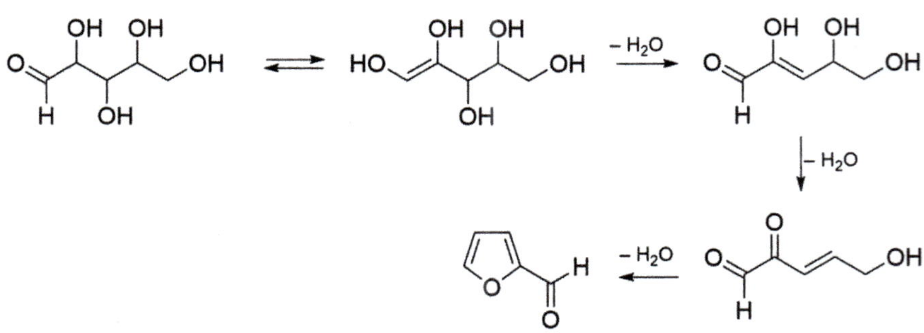

Figure 2.17 The conversion of C5 sugars to furfural.

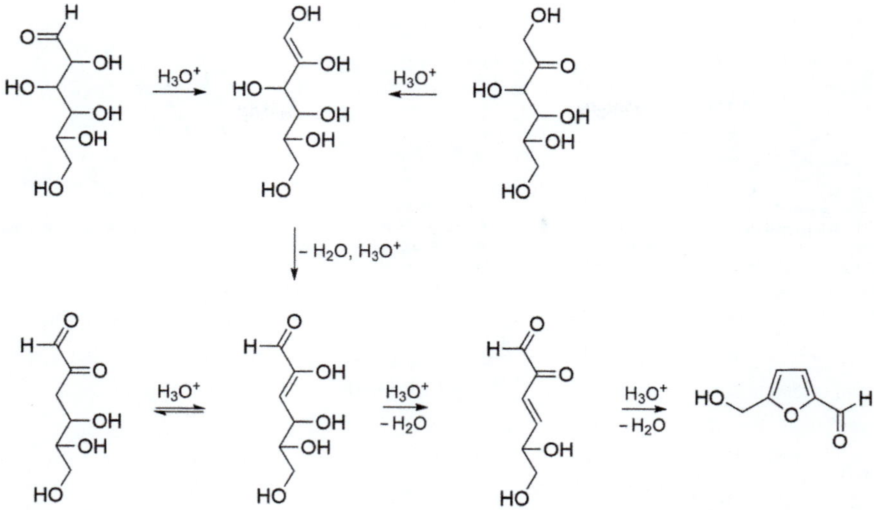

Figure 2.18 The conversion of C6 sugars to hydroxymethylfurfural.

option, because the diacid, together with its dichloride and its esters, is the most important monomer derived from HMF, as described in detail in the polycondensation chapter.

In more general terms, the role of monomer precursor of HMF represented the fundamental qualitative change in furan polymer chemistry, since its double substitution with reactive moieties (as opposed to the singly substitute F), opened a large avenue of possibilities related to preparing specific 2,5-disubstituted structures suitable for step-growth processes.

Furan polymers must be approached in the first approximation as the outcome of two fundamental building blocks based on renewable resources, namely F and HMF.

A rich variety of monofunctional monomers, outlined in Figure 2.19, have been prepared from F, and their polymerizations, as well as the properties of the ensuing materials, are one of the fundamental topics of this book. Apart from furfuryl alcohol, discussed in a separate chapter, all these monomers polymerize or copolymerize through the function external to the furan heterocycle by chain reactions, thus producing the generic macromolecular structure shown in Figure 2.20, where the ring is appended to it as a side group. The same applies to the corresponding branched and cross-linked architectures.

The difunctional monomers synthesized from HMF are given in Figure 2.21 and their polymerizations by stepwise reactions constitute, together with the properties of the ensuing materials, the second fundamental topic of this book. The generic macromolecular structure arising from their polymerization and copolymerization is depicted in Figure 2.22, where the furan ring is now an integral

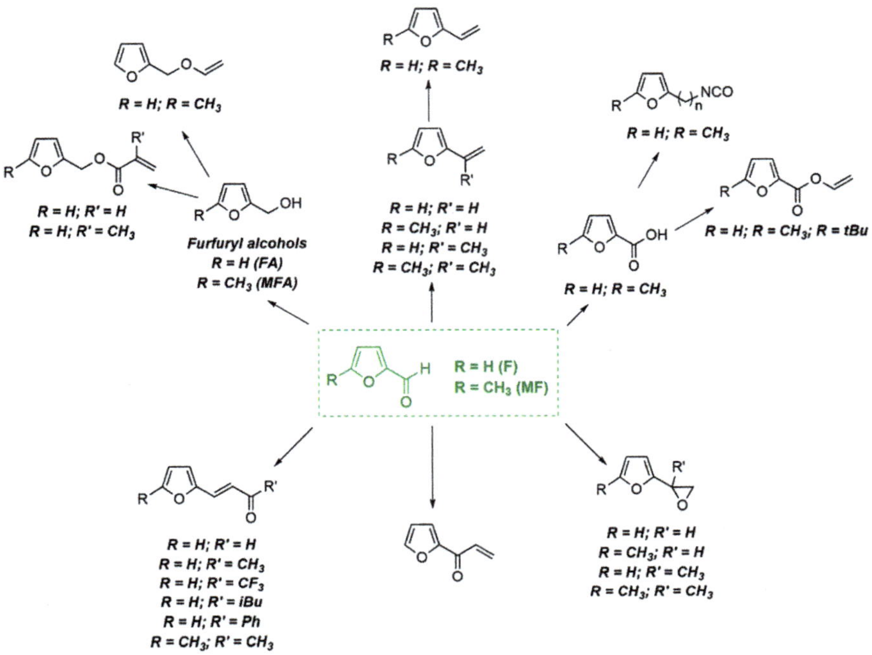

Figure 2.19 The most important monomers synthesized from F or 5MF.

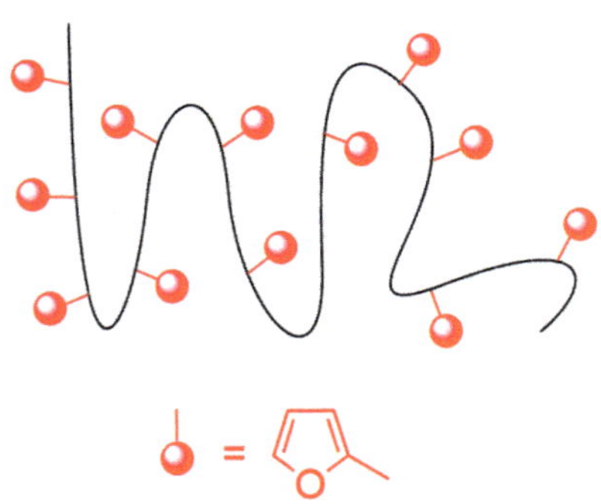

Figure 2.20 The general structure of polymers and copolymers prepared from the monomers shown in Figure 2.19 (except for furfuryl alcohol).

part of the chain, viz. it is incorporated in the polymer backbone. Again, this also applies to branched and cross-linked macromolecular architectures.

An alternative approach to prepare bifunctional monomers was developed before the irruption of HMF on the scene and is based on an acid-catalyzed coupling reaction involving the condensation of two monosubstituted furan molecules and an aldehyde or ketone, as shown in Figure 2.23 [17]. More recently, the

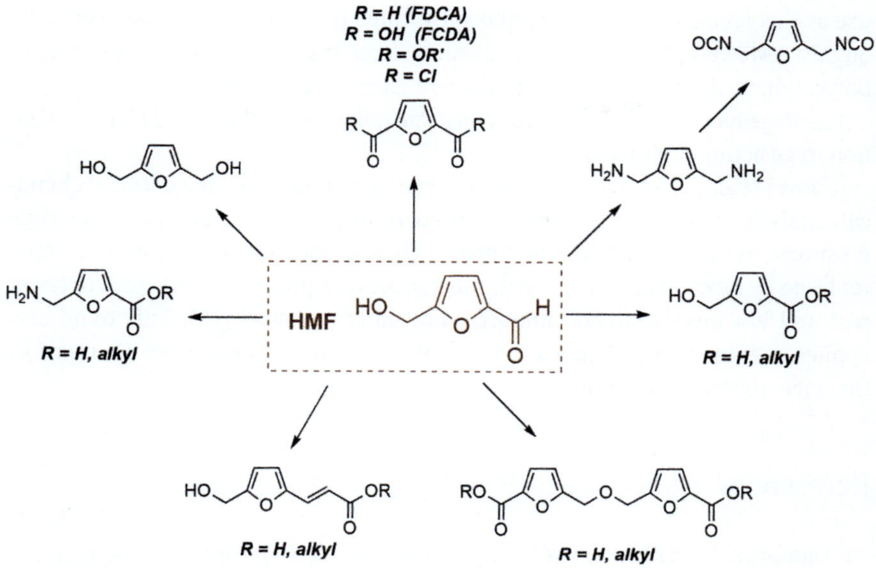

Figure 2.21 The most important monomers synthesized from HMF.

Figure 2.22 The generic structure of polymers and copolymers derived from the monomers shown in Figure 2.21.

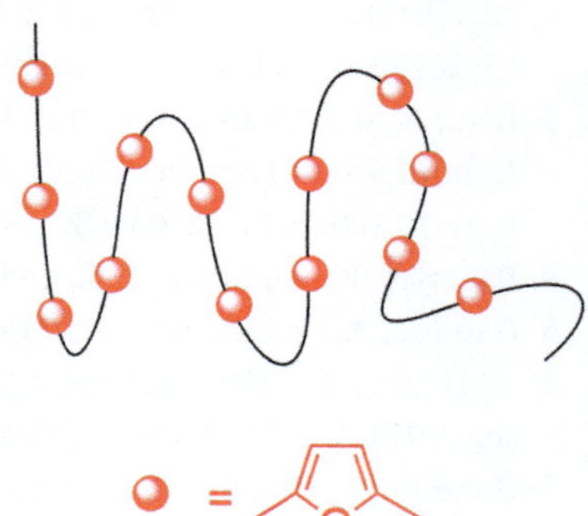

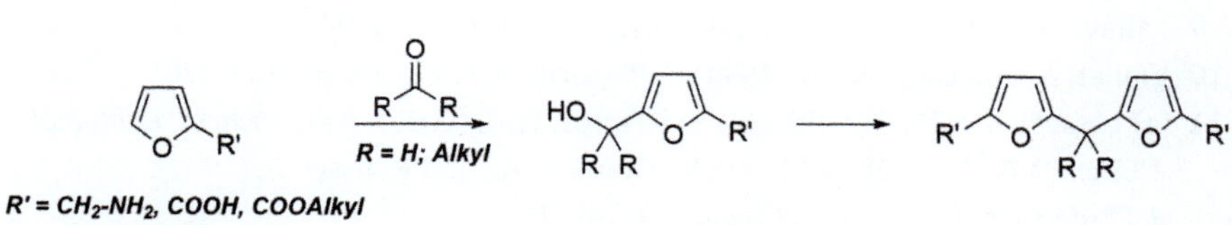

Figure 2.23 Electrophilic condensation leading to difuran monomers for stepwise polymerizations.

use of F as precursor to a variety of disubstituted furan monomers has been thoroughly reviewed [18]. Most of the ensuing monomers were used in a variety of polycondensations, as discussed in the corresponding chapter.

The stage has been set for the various furan polymer synthesis and characterization to be acting their roles.

A novel research area is currently developing in which the conventional chemical catalytic synthetic procedures to prepare furan derivatives from renewable resources, notably F, HMF and some of their monomer derivatives, are being replaced by biocatalytic and fermentative processes, proposing processes of lower cost and low environmental impact, as recently reviewed [19]. This trend also applies to enzymatic polymerization involving furan monomers, as discussed in the polycondensation section.

References

1 Dunlop, A.P. and Peters, F.N. (1953). *The Furans. ACS Monograph Series* 119. Reinhold Publishing Co., New York.

2 (a) Keay, B.A. and Dibble, P.W. (1997). In: *Comprehensive Heterocyclic Chemistry II*, Vol. 2. (eds. A.R. Katritsky, C.W. Rees, and E.F.W. Scriven). Elsevier, Oxford, 395. (b) Hou, X.L., Yang, Z., and Wong, H.N.C. (2003). *Progr. Heterocycl. Chem.* 15: 167–205. (c) Duc, D.X. (2019) *Mini-Rev. Org. Chem.* 16: 422–452.

3 Boto, A. and Alvarez, L. (2011). Furan and Its Derivatives. In: *Heterocycles in Natural Product Synthesis*. (eds. K.C. Majumdar and S.K. Chattopadhyay). DOI: 10.1002/9783527634880.ch4.

4 Banerjee, R., Kumar, H.K.S., and Banerjee, M. (2012). *Int. J. Rev. Life. Sci.* 2: 7–16.

5 Gandini, A. and Rieumont, J. (1976). *Tetrahedron Lett.* 25: 2101–2104.

6 (a) D'Auria, M. (1997). *Internet J. Sci.* 4. Available online: http://www.netsci-journal.com/97v4/97015/index.htm. (b) D'Auria, M. (1999). *Heterocyles* 50: 1115–1136.

7 Gandini, A., Parsons, J.M., and Back, R.A. (1976). *Canadian J. Chem.* 54: 3095–3101.

8 Galego, N. and Gandini, A. (1977). *J. Polym. Sci. Polym. Chem. Ed.* 15: 1027–1028.

9 Lahav, M. and Schmidt, G.M. (1967). *J. Chem. Soc. B* 239–245.

10 Viallet, A. and Gandini, A. (1990). *J. Photochem. Photobiol.* 54: 129–130.

11 (a) Lasseguette, E., Gandini, A., and Timpe, H.-J. (2005). *J. Photochem. Photobiol. A Chem.* 174: 222–228. (b) Baret, V., Gandini, A., and Rousset, E. (1997). *J. Photochem. Photobiol. A Chem.* 103: 169–175.

12 (a) Taticchi, F. and Fringuelli, A. (Eds.) (1990). *Dienes in the Diels–Alder Reaction*. Wiley-Interscience, New York, 368. (b) Taticchi, F. and Fringuelli, A. (Eds.) (2002). *The Diels–Alder Reaction: Selected Practical Methods*. John Wiley & Sons, New York, 356.

13 Froidevaux, V., Borne, M., Laborbe, E., et al. (2015). *RSC Adv.* 5: 37742–37754.

14 Cioc, R.C., Crockatt, M., Van der Waal, J.C., and Bruijnincx, P.C.A. (2022). *Angew. Chem. Int. Ed.* 61: e202114720.

15 (a) Li, X., Jia, P. and Wang, T. (2016). *ACS Catal.* 6: 7621–7640. (b) Machado, G., Leon, S., Santos, F., et al. (2016). *Nat. Resour.* 7: 115–129. (c) Delbecq, F., Wang, Y., Muralidhara, A., El Ouardi, K., Marlair, G. and Len, C. et al. (2018). *Front. Chem.* 6: Article 146. DOI: 10.3389/fchem.2018.00146.

16 (a) van Putten, R.-J., van der Waal, J.C., de Jong, E., et al. (2013). *Chem. Rev.* 113: 1499–1597. (b) Rosatella, A.A., Simeonov, S.P., Frade, R.F.M., and Afonso, C.A.M. (2011). *Green Chem.* 13: 754–793. (c) Thoma, C., Konnerth, J., Sailer-Kronlachner, W., et al. (2020). *ChemSusChem* 13: 3544–3564. (d) Messori, A., Fasolini, A., and Mazzoni, R. (2022). *ChemSusChem* 15: e202200228.

17 Pennanen, S. and Nyman, G. (1972). *Acta Chem. Scand.* 26: 1018–1022.

18 Zhu, J. and Yin, G. (2021). *ACS Catal.* 11: 10058–10083.

19 (a) Lalanne, L., Nyanhongo, G.S., Guebitz, G.M., and Pellis, A. (2021). *Biotechnol. Adv.* 48: 107707. (b) Becerra, M.L., Prieto, G.A., Rendueles, M., and Diaz, M. (2022). *Biomass Conv. Bioref.* DOI: 10.1007/s13399-022-03535-5.

3

Polymers from Furfural and Furfuryl Alcohol

3.1 Furfural Resins

As already mentioned, the vast majority of F industrial output is channeled to produce FA. The remaining uses include extraction for aromatics from lubricating oils, purification solvent for C4 and C5 hydrocarbons, reactive solvent and wetting agent, chemical feedstock for furan derivatives other than FA, nematode control and monomer for thermoset resins [1], which is the only topic treated here.

When highly purified F is heated *in vacuo* between 100 and 200 °C, it develops a yellow color and a black resin deposits on the vessel's walls [2], the initial manifestation of these phenomena being very temperature dependent, e.g., one week at 100 °C and a few hours at 200 °C. Workup of the reaction products from an experiment with 200 g of F, gave very small yields of a white crystalline material (0.05 g) and of an yellow oily liquid (0.1 g), as well as 5 g of a resin of which only 0.3 g were soluble in organic solvents. The structure of the crystalline product was assessed spectroscopically as shown in Figure 3.1, which also puts forward the mechanism of its formation, namely an oxidation/reduction involving three F molecules. The soluble resin was identified as a trifuran derivative arising from the condensation of three F molecules, as shown in Figure 3.2. As for the cross-linked black resin, elemental analysis and FTIR spectroscopy suggested the tentative structure sketched in Figure 3.3, arising from those two precursors growing through a free radical mechanism. Interestingly, studies on the acid-catalyzed pyrolysis of furfural, suggested that the ensuing resins were generated by similar mechanisms and bore likewise cross-linked structures [3, 4].

The resins prepared from F alone have not found significant technological applications, but the use of F as a comonomer has provided resins with a somewhat more successful outcome, albeit not very widespread. Thus, the phenol-F compositions have been described in the elaboration of wood adhesives [1c] and

Furan Polymers and their Reactions, First Edition. Alessandro Gandini and Talita Martins Lacerda.

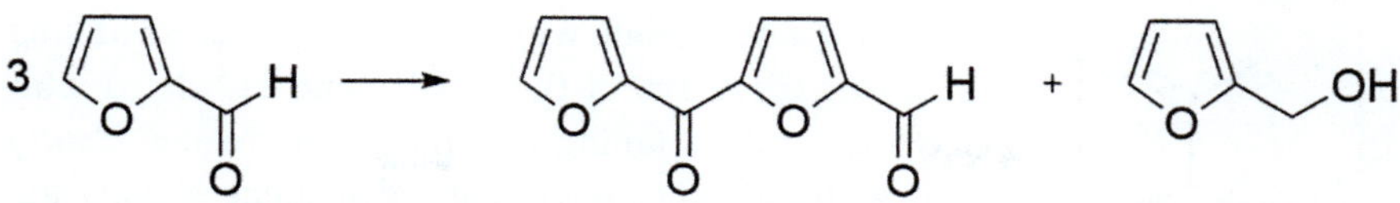

Figure 3.1 The crystalline difuran compound formed in the thermal treatment of F.

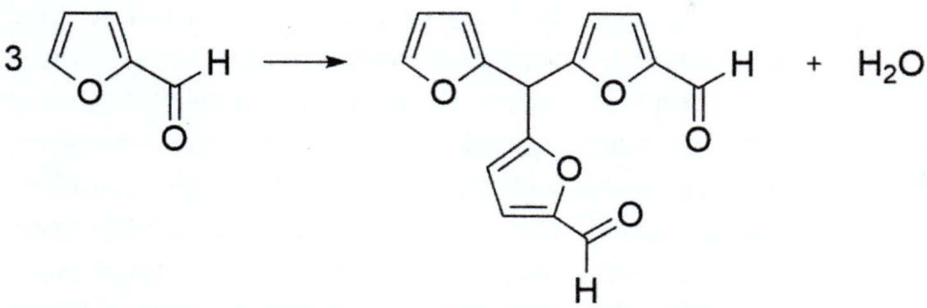

Figure 3.2 The oily trifuran compound formed in the thermal treatment of F.

protective coatings [5] and composites with clay [6] or soybean oil [7]. However, the most interesting studies in this context concern the use of the resinification of the reaction products between F and acetone [8], viz. the aldol-crotonic condensation given in Figure 3.4, where the compound with R'=H and R=CH_3, furfurylidene methyl ketone, was submitted to an acid-promoted polymerization. The initial polymer structure is mostly linear, as shown in Figure 3.5, but soon electrophilic reactions involving the carbonyl function and the furan C5 position give rise to cross-linked resinous products, which have found applications as adhesives for abrasive materials and composites with clays and are commercialized in several countries. When two molecules of furfural are reacted with acetone, the corresponding difurfurylidene ketone is obtained, whose acid-catalyzed resinification proceeds as with its

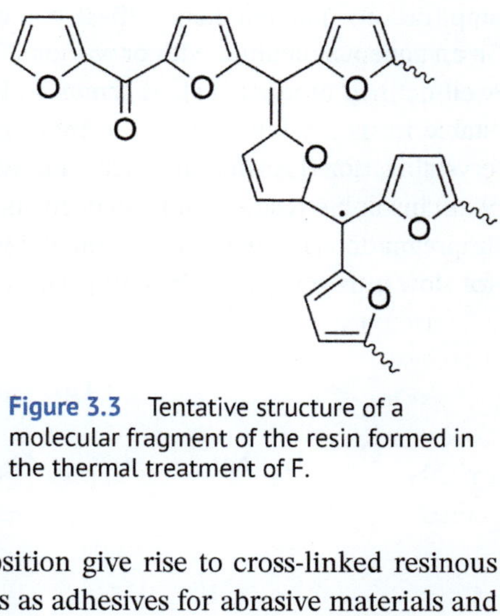

Figure 3.3 Tentative structure of a molecular fragment of the resin formed in the thermal treatment of F.

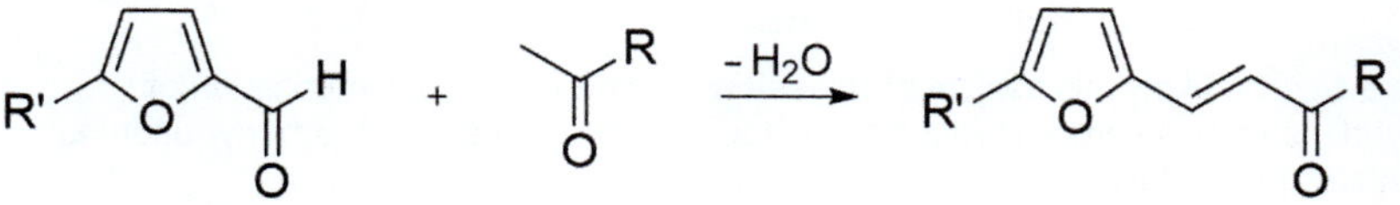

Figure 3.4 The base-catalyzed condensation of F and MF with ketones.

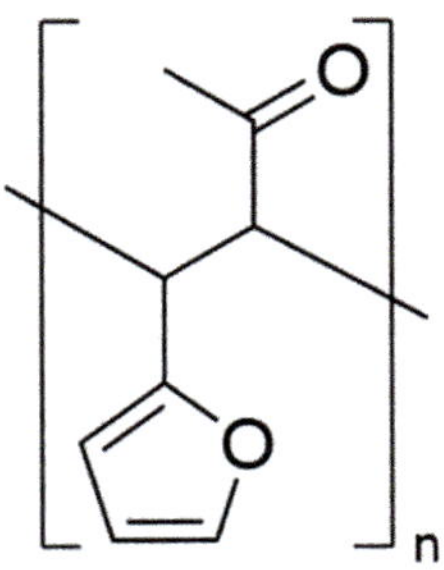

Figure 3.5 The early linear structure of the furfurylidene methyl ketone polymerization.

mono-homologue, but with a more rapid cross-linking [9]. The dark colors of these polymers, associated with electronic spectra tailing into the visible region, clearly suggest the presence of conjugated sequences along the macromolecular chains.

Furfurylidene methyl ketone has also been submitted to anionic polymerization [10], using different nucleophilic initiators and a number of model compounds, and found to undergo a process involving an isomerization step previous to chain growth, as shown in Figure 3.6, but some propagation through the opening of the carbonyl moiety was also found to occur. The ensuing complex resin structure is therefore not conducive for a well-defined material, although some commercial products prepared by the anionic polymerization are commercially available for use as adhesives.

Resins based on the F-urea combination have attracted attention in several applications. The reaction of these two compounds has been thoroughly studied in an aqueous medium with or without HCl as catalyst, and its first step leads to a well-defined product [11], as shown in Figure 3.7. Although this intermediate is stable in its crystalline form, in more general stoichiometric conditions where crystallization does not take place, the system is known to evolve progressively to yield insoluble resins that have been successfully applied to wood adhesion and impregnation [12], to replace formaldehyde in urea-formaldehyde resins [13], and for slow nitrogen release in soils [11].

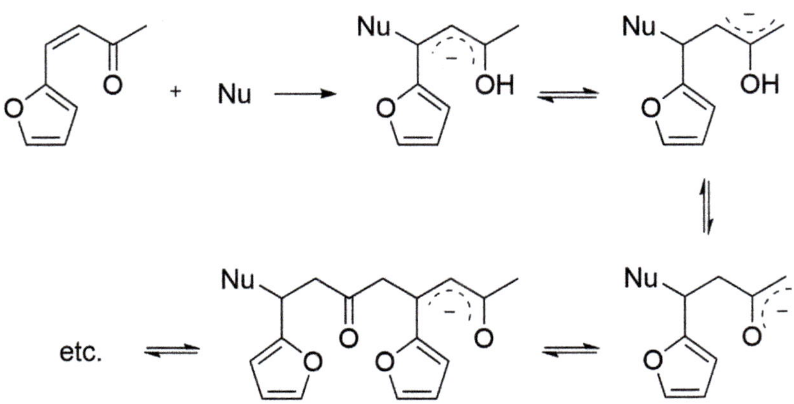

Figure 3.6 The likely mechanism of the anionic isomerization polymerization of furfurylidene methyl ketone through the external alkene moiety, but carbonyl-opening growth also takes place.

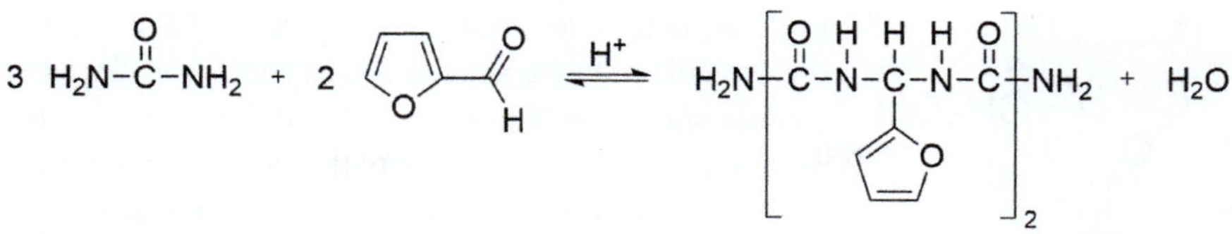

Figure 3.7 The optimized condition for the F + urea to give a well-defined compound.

On the whole F-based resins do not represent an important technological niche, compared with their FA counterparts, but their role as replacements of the harmful formaldehyde should see a more sustainable development in the future.

3.2 Furfuryl Alcohol Resins

Furfuryl alcohol (FA) is still today by far the most important furan compound derived from F. The industrial reduction process (Figure 3.8) is carried out predominantly using a copper catalyst operating in the gas or the liquid phase. Numerous other processes have been studied, including a green approach using enzymes, and a direct synthesis from xylose [14]. FA finds application mostly in the fabrication of resins, as discussed here, but also as a lubricant, a dispersing agent, in food production, as well as being an important intermediate in fine chemicals synthesis like lysine, vitamin C and levulinic acid. The FA market, estimated at EUR 500 million in 2019, is expected to triple by 2028, mostly for its use in the polymers and adhesives sector, as a solvent and wetting agent. This section is exclusively devoted to the FA macromolecular materials.

3.2.1 The Acid-catalyzed Polycondensation of Furfuryl Alcohol

From the very beginning of the exploitation of FA as a source of resinous materials, mostly under acidic conditions [1a], strong scientific and technical interest attracted researchers to unravel the mechanisms leading to these black insoluble polymers, which however displayed very promising properties in several domains of applications that have not stopped widening. In order to set the stage for this

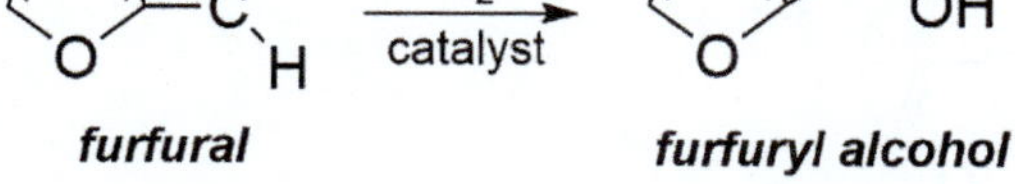

Figure 3.8 The catalyzed hydrogenation of F to FA.

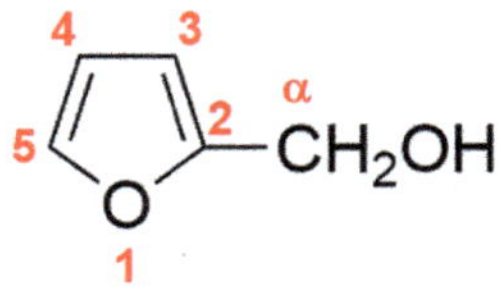

Figure 3.9 FA seen as a compound with several potential reactive sites.

lively and often contentious saga, Figure 3.9 illustrates the possible reactive sites of FA. When FA is placed in contact with an acidic reagent, it starts reacting, as first revealed by a progressively bathochromic coloration, and then by yielding a dark insoluble resin, that eventually turns into a black cross-linked material. This qualitative behavior is entirely "universal" in the sense that such parameters as the strength of the acid, the temperature and the reaction medium, do not affect the occurrence of these events until the ultimate result, but of course can strongly determine their kinetic flow, which can take seconds or months to finalize.

We will not approach this issue historically because whereas some studies were able to shine some light on the chemical intricacies of the system, many others failed to provide much advance in their understanding. We prefer therefore to go straight to what was the unraveling of the main puzzles and work back to refute, with what we believe are convincing arguments, previous and later suggested alternative mechanisms. The investigation in question [15] called upon a systematic approach requiring a large number of FA homologues and other compounds and reagents, in order to tackle as many possible mechanistic steps as possible. The basic tenant of FA acid-catalyzed polycondensation stems from two well-recognized alternative initial steps, as displayed in Figure 3.10.

Given the ease of formaldehyde loss arising from the head-to-head coupling, the net result is in fact practically the same as with the head-to-tail condensation, in the sense that the pursuit of these steps should lead to the linear colorless poly(2,5-furyl methylene) shown in Figure 3.11, if no other reaction intervenes to perturb this simple macromolecular growth. Should the formaldehyde loss be incomplete during this sequence of condensations, the ensuing polymer will still be similarly linear and colorless, but it will incorporate some dimethylene oxide ether moieties in its chains. It would also be expected that this attractive furan material will be thermoplastic and soluble in a number of solvents.

The harsh reality is however that, despite the established initial growth mechanism of Figure 3.10, events start degenerating soon afterwards with color

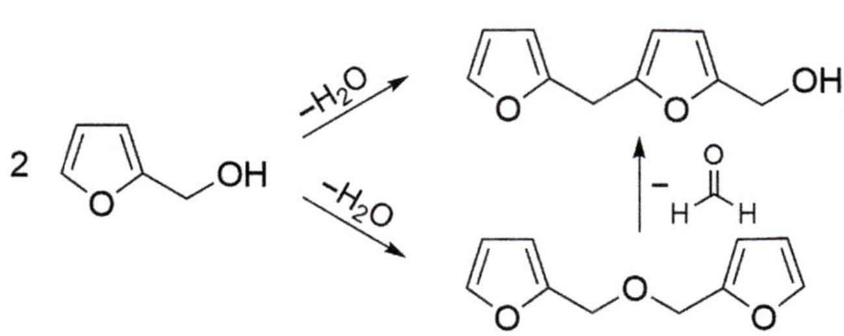

Figure 3.10 Two alternative condensation steps of FA catalyzed by acids.

development and ultimate precipitation of a black cross-linked product. This disappointing and initially obscure situation spurred a host of studies throughout the second half of the twentieth century, but it took a more thorough approach to unravel these intricacies [15]. The color problem must be discussed in the first place, since it precedes, and indeed is responsible, for the subsequent cross-linking. The main culprit is the methylene bridge joining the two furan rings at their C2 and C5 positions, because of the lability of the C-H bond caused by its direct attachment to the dienic heterocycle, which facilitates the loss of a hydride ion induced by the carbocationic growing species. Interestingly, this reaction was also observed to occur when FA was replaced by its α-methyl homologue, which produced the same overall phenomenology during its polycondensation, since it still bore one mobile hydrogen atom, whereas the dimethyl counterpart behaved in a totally different manner in that it polymerized smoothly, albeit slowly because of steric hindrance, to yield the linear structure given in Figure 3.12 without either color development or cross-linking [15].

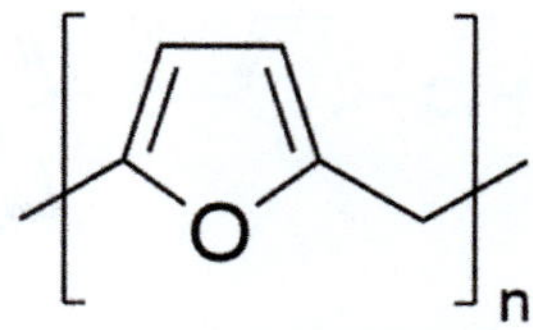

Figure 3.11 The ideal polymer structure arising from the succession of the condensation reactions of Figure 3.10.

This hypothesis was further confirmed by using polymer model compounds reacting with hydride-ion abstractors [15], as shown in Figure 3.13 for dimers and Figure 3.14 for the FA-polymer trimer model, which also includes the carbocation neutralization to give a novel trimer incorporating a central 2,5-dihyrofuran moiety with 2,5-exocyclic unsaturation conjugated with the two side furan rings.

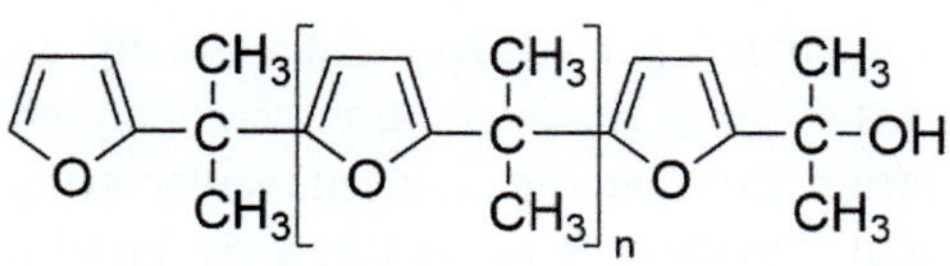

Figure 3.12 The structure of the colorless linear polymer formed in the acid-catalyzed polycondensation of α,α-dimethylFA.

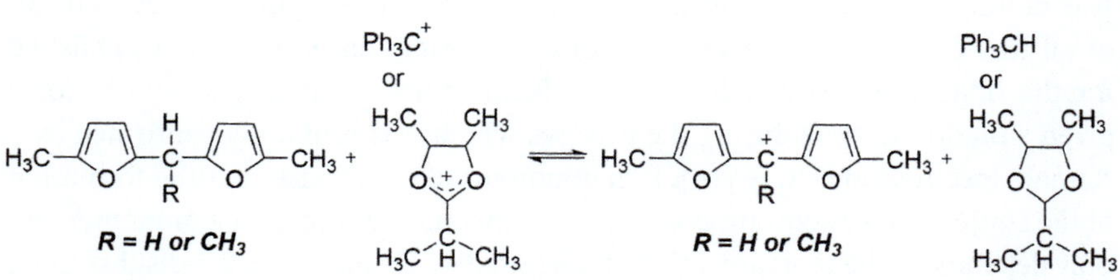

Figure 3.13 The reaction of dimer model compounds of FA and α-methylFA with hydride ion abstractors.

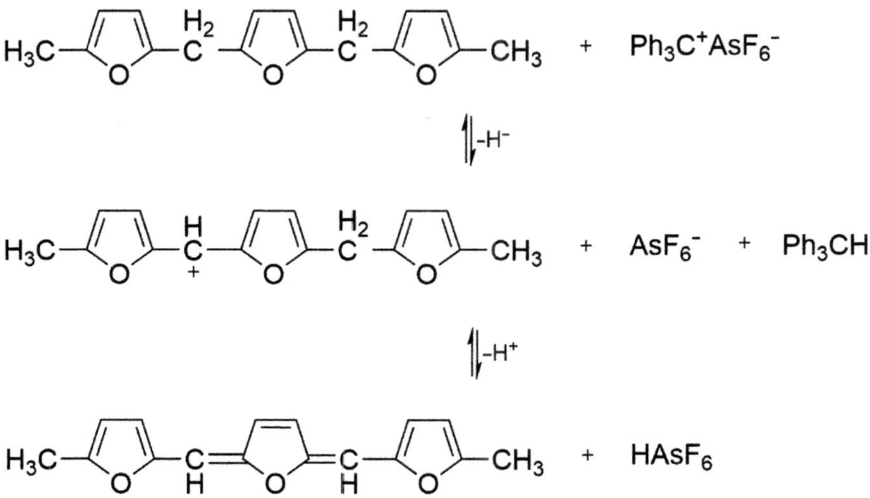

Figure 3.14 The reaction of FA-polymer trimer model with a hydride ion abstraction followed by the neutralization of the ensuing carbocation.

These reactions, followed by neutralization with trimethylamine, were also monitored by UV-visible spectroscopy, which confirmed the formation of both the intermediate carbocations (strong bathochromic shift from 260 nm to 500 nm for the dimer and 600 nm for the trimer) and the neutralized conjugated structure, which gave a maximum at 470 nm. The α,α-dimethylFA models did not react either with the hydride-ion abstractors, or any acid initiators, thus confirming once more the validity of the mobile-hydrogen mechanism as responsible for color formation during both FA and α-methylFA polycondensations. Finally, an FA oligomer, isolated in the early stages of the polycondensation was analyzed spectroscopically and its ^{1}H NMR spectrum displayed the expected regular structure with an additional resonance attributed to the presence of C2 methyl groups, as indicated in Figure 3.15. Its FTIR spectrum confirmed the expected structure and also gave a weak carbonyl peak at 1715 cm^{-1} (see later discussion). The treatment of this oligomer with the hydride-ion abstractor, followed by neutralization, was monitored by UV-visible spectroscopy, as shown in Figure 3.16. As a result of all this evidence, the mechanism of color formation was clearly established for the first time [15] and its essential features are given in Figure 3.17 for a given situation in time during the process, which was brilliantly confirmed by a Raman-spectroscopy study [16]. It is important to emphasize that the formation of the conjugated systems proceeds without the intervention of the monomer and only depends on the reorganization of the initial $-CH_2$-Fu- (Fu = 2,5-disubstituted furan) sequences in the formed polymer under the effect of the acid medium. In other words, the process in Figure 3.17 can continue as long as these sequences

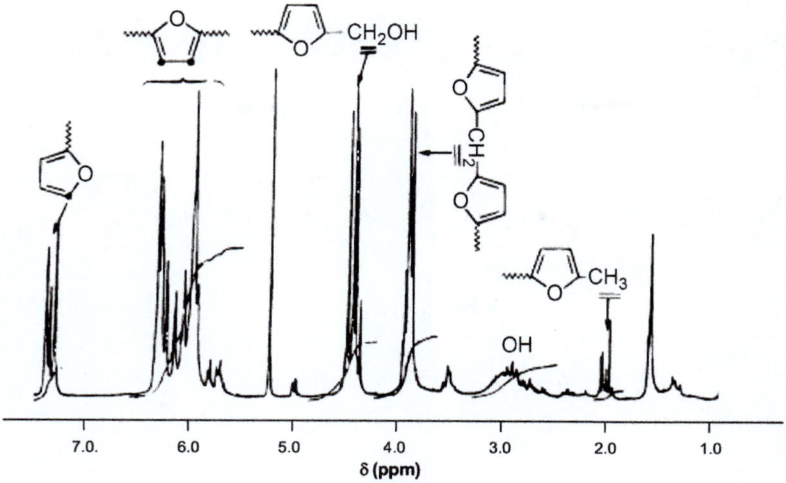

Figure 3.15 A typical ^{1}H NMR spectrum of an FA oligomer isolated in the early stages of the polycondensation.

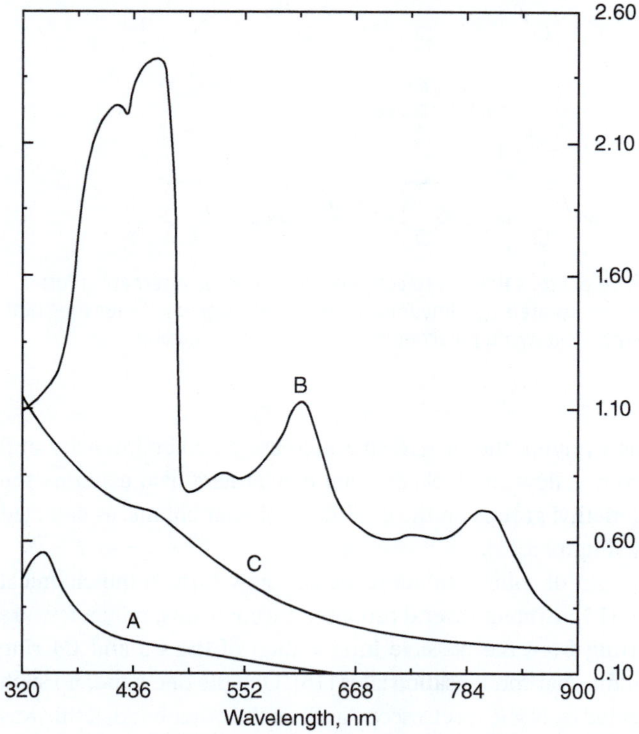

Figure 3.16 UV-visible spectra of (A) the FA oligomer, (B) after its reaction with the hydride-ion abstractor, and (C) after neutralization with trimethylamine.

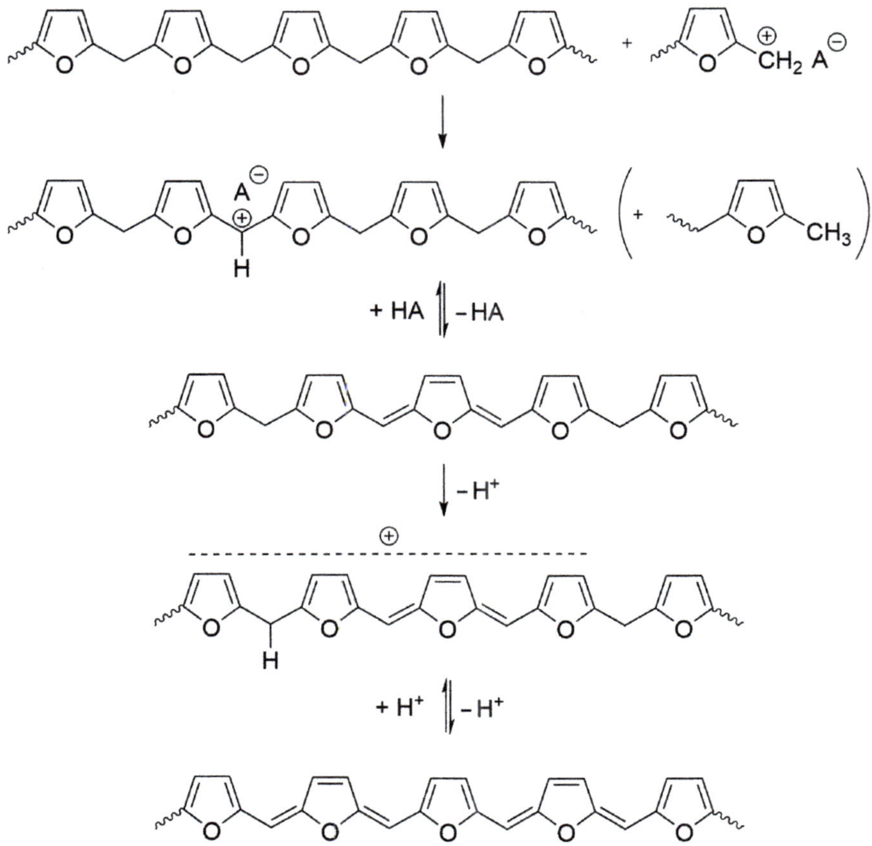

Figure 3.17 The sequence of steps leading to conjugated sequences alternating furan heterocycles and 2,5-exounsaturated 2,5-dihydrofuran moieties, responsible for the color development during the acid-catalyzed polycondensation of furfuryl alcohol.

are available, and this explains the progressive increase in color intensity and bathochromic shift from yellow to blackish. This mechanism also explains the formation of terminal methyl groups on the modified polymer chains, as detected by NMR spectroscopy (Figure 3.15).

Having solved the origin of color formation, we can now turn to the chemical source of cross-linking [15]. Among several proposed mechanisms, only a few are worth discussing, starting from the possible intervention of the C3 and C4 ring positions as possible additional condensation sites [16]. One the one hand, no such evidence was ever detected by NMR spectroscopy and, on the other hand, if this was the case, α,α-dimethylFA should have ultimately yielded a cross-linked polymer, which was not the result of its polycondensation that gave a linear polymer without any evidence of the intervention of the C3 and C4 positions [17]. Another suggestion called upon the role of the formed water as a source of ring hydrolysis and subsequent branching reactions induced by the ensuing aliphatic products. This hypothesis was brilliantly refuted by a study of the polycondensation of furfuryl acetate [18], which gave the same overall phenomenology of that of FA in the absence of water as condensation product, here replaced by acetic acid. In fact, the presence of generated moisture in the FA system does play a minor role as a source of hydrolysis, indicated by the presence of weak carbonyl groups in the FTIR spectra of FA polymers, but it is definitely a negligible side reaction and certainly not a cause of any detectable polymer branching. A more challenging idea invoked the methylene bridges in the polymer as reactive sites of condensation either with a growing FA -OH group, or with the formaldehyde released in the head-to-head couplings [19]. These interesting hypotheses were however refuted by the two model experiments shown in Figure 3.18, which both gave negative outcomes, thus ruling out the corresponding branching mechanisms [15].

An important point in this context is the fact that no evidence of cross-linking is observed during the "normal" polymer growth and manifests itself after the color development has proceeded to a high degree of advancement, suggesting that the conjugated sequences are indeed needed for this event to occur. It is also uncommon that the viscosity of the medium does not increase in any appreciable way as before the polymer precipitation, i.e., that the approach to cross-linking is not accompanied by the classical dramatic increase in viscosity. The observation of the two main moieties in the highly conjugated polymers, which have modest DP

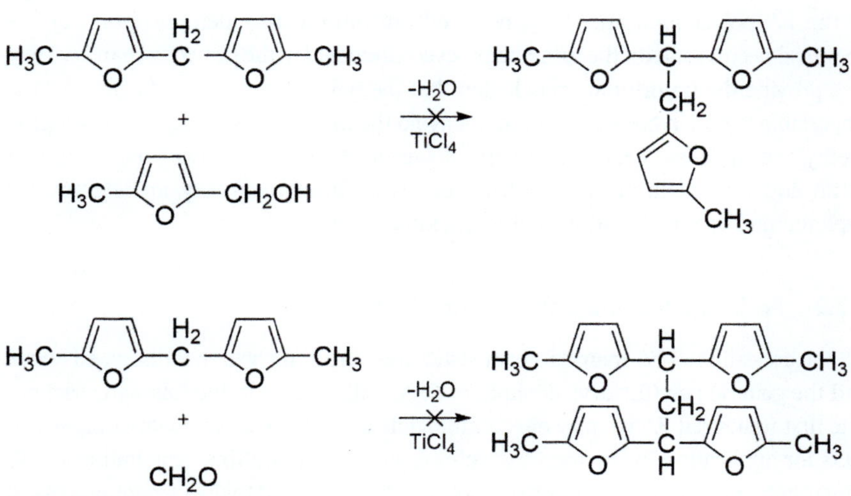

Figure 3.18 Two model experiments showing that the polyFA methylene bridges are not responsible for branching reactions.

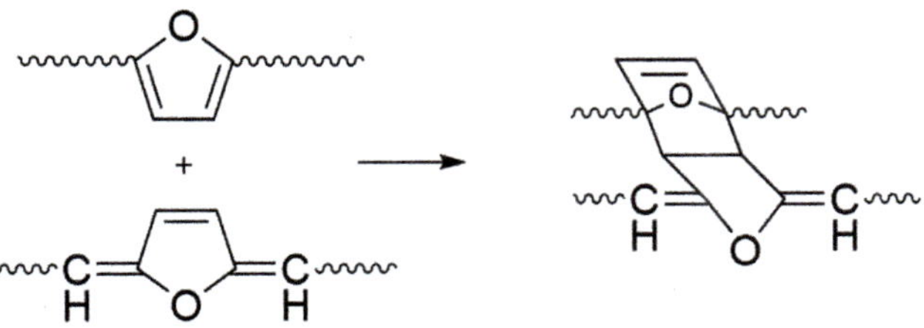

Figure 3.19 The DA coupling reaction responsible for polymer-to-polymer cross-linking.

values of 10–20 before precipitation, suggests the possibility that their DA coupling between furan diene units and 2,5-exounsaturated dihydrofuran dienophile units is the cause of cross-linking, as shown in Figure 3.19 for a single event, which of course is repeated several times among the chains [15].

A comprehensive study of the acid-catalyzed polycondensation of 2-thienyl methanol (TM) and of its variously substituted compounds [20] showed two major differences in behavior with respect to that of its FA homologues, namely that, on the one hand, the C3 and C4 positions are frequently involved in condensation reactions together with the expected C5 sites, and, on the other hand, the polymers do not cross-link. The TM polymers give rise to the same sequences of conjugated structures, accompanied by color formation; but since the thiophene heterocycle does not display an adequate diene character, interchain DA coupling is not possible and hence the lack of cross-linking. For the same reasons, the electrophilic substitutions are now not limited to the C2-C5 regiospecificity displayed by the furan ring with the FA polycondensation.

The mechanistic features that finally rationalized the intricate phenomenology of the FA polycondensation [15] represent an illuminating demonstration of the chemical peculiarities of the furan heterocycle discussed in the previous chapter. They also provide the unfortunate conclusion that the polymer structure of Figure 3.11 is impossible to obtain because it could not avoid the intrinsic instability of the bridging methylene moieties. FA oligomers are however commercially available, including in an aqueous medium, notably from the Transfurans company in Belgium, for exploitation in the elaboration of different kinds of resins.

3.2.2 FA-based Materials and Applications

All the cross-linked FA materials bear structures very similar to that discussed above and the generic poly(furfuryl alcohol), or PFA, will be used in the following section. The first industrial FA resins were commercialized by the Quaker Oats Company in 1958 for application in the foundry technologies [1,21] and this exploitation is still very much in use and includes several compositions adapted to the present processes. The other major outlets for FA-based resins are in the refractory and anticorrosion domains, in addition to a large variety of additional, more specific areas such as metal casting cores and molds, corrosion-resistant coatings, polymer concretes, wood adhesives and binders, sand consolidation and well plugging, low flammability and low smoke release materials and carbonaceous products, including graphitic electrodes and carbon micro-particles [1, 21]. More recently, the combined use of FA and another renewable resource stands out as particularly promising both in terms of the properties of the materials and for the obvious sustainability value. Thus, tannins were copolymerized with FA to prepare thermosets that included covalent bridges between the two polymers, displayed Tg values above 200 °C and remarkable resistance to mechanical ware, hence highly suited for such applications as grinding wheels [22] and automotive brake pads [23]. The other interesting marriage involves FA and lignin [24, 25], where again strong NMR evidence proved the formation of covalent links during the growth of the co-cross-linked materials through the electrophilic substitution involving the poly-FA growing carbocation and an unsubstituted position of a lignin aromatic ring, as shown in Figure 3.20 [25]. In a similar, more technical study FA was copolymerized with lignosulfonates and tannin to prepare adhesives for solid wood or particle boards in substitution of the less ecological phenol-based counterparts [26].

In a different and more significant vein, the impregnation of wood with FA followed by its *in situ* resinification, has received considerable attention in many countries and reached industrial production. The cell walls of the wood are swollen with furfuryl alcohol, which is then polymerized within the cell walls using latent acidic catalysts. With this treatment, the water absorption capacity of the wood is drastically reduced and the tendency for wood to deform when there is a change in humidity is minimized. The hardness and rot resistance of the modified wood is also improved compared to that of native wood. This area of investigation became attractive in the early years of the millennium, when a seminal study in 2004 [27] set the pace for a rapid development, although previous reports had described its potential. The most frequently adopted catalyst is maleic anhydride, but a mixture of oxalic and citric acid proved equally successful, as well as other co-reactants, while reducing the amount of FA needed for the optimized impregnation [28]. The protection of wood furfurylation against marine boring crustacean is another major advantage [29], since it allows the use of the treated material in prolonged sea-submersion conditions. In all these systems, the growing FA polymer is most

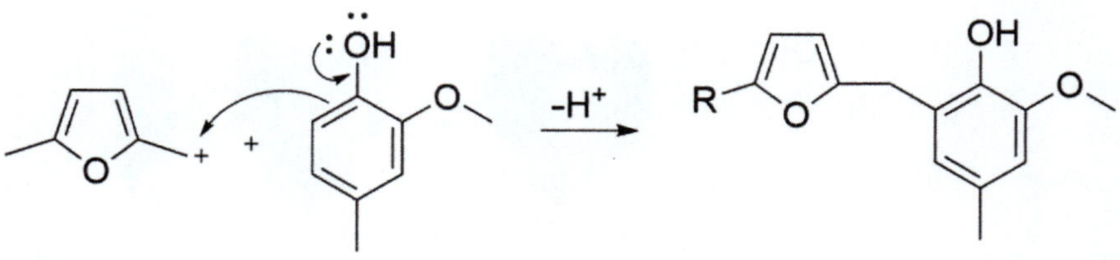

Figure 3.20 The coupling of a growing poly-FA chain with lignin.

likely to react with the wood lignin following the mechanism given in Figure 3.20 [25, 27]. A recent addition to this topic [30] describes the treatment of poplar wood by impregnating only one side of panels with FA and thereafter inducing its polymerization. The purpose of this approach was to improve the dimensional stability and mechanical properties of the panels intended to be used for construction and interior decoration applications. The ongoing advances on wood modification with FA have inspired their extension to the use of other non-toxic renewable resources like rosin, plant oils and biodegradable polymers as impregnating agents [31].

FA-based rigid foams, bioplastics, wood- and bio-adhesives have been thoroughly reviewed [32]. By far the most interesting among these materials, and indeed those which have not been discussed here previously, are furan–tannin foams and stiff materials, which can be obtained by different techniques and compositions (i.e., the proportion of FA, the type of tannin, and the presence of a third property-specific component), in order to prepare both rigid carbonaceous materials, ideal for high-temperature insulation, and flexible counterparts for a wider range of applications [32]. NMR and MALDI-ToF evidence clearly proved that the two growing polymers did indeed marry by mutual covalent bonding to give a co-continuous percolating network. Figure 3.21 illustrates rigid objects obtained by this renewable-resource copolymerization.

A novel strategy, based on the cationic polymerization of FA with tung oil main fatty acid, i.e., once again two monomers from renewable resources, has been implemented with the purpose of preparing co-cross-linked materials possessing intermediate thermal and mechanical properties between those of rigid PFA and of low-Tg oil-based network. In a first study, methyl-α-eleostearate plus FA, or furfuryl-α-eleostearate were treated with trifluoroacetic acid and the progress of

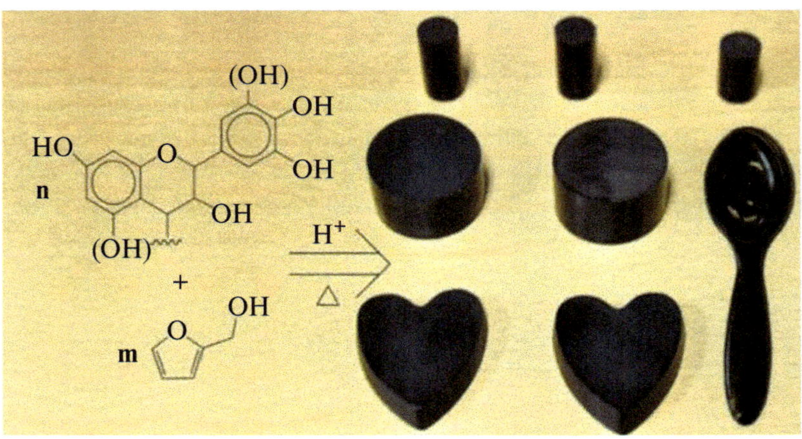

Figure 3.21 Rigid artefacts prepared by the copolymerization of FA and tannin. From ref.(32)., 2021 / Reproduced with permission from IntechOpen / CC BY 3.0.

the reaction was followed by ^{1}H NMR spectroscopy until the formation of gelled products, which were thoroughly characterized [33] to prove that their structures were indeed made up of covalently-joined networks. Figure 3.22 shows a view of the process involving the first system. In order to compensate for the different

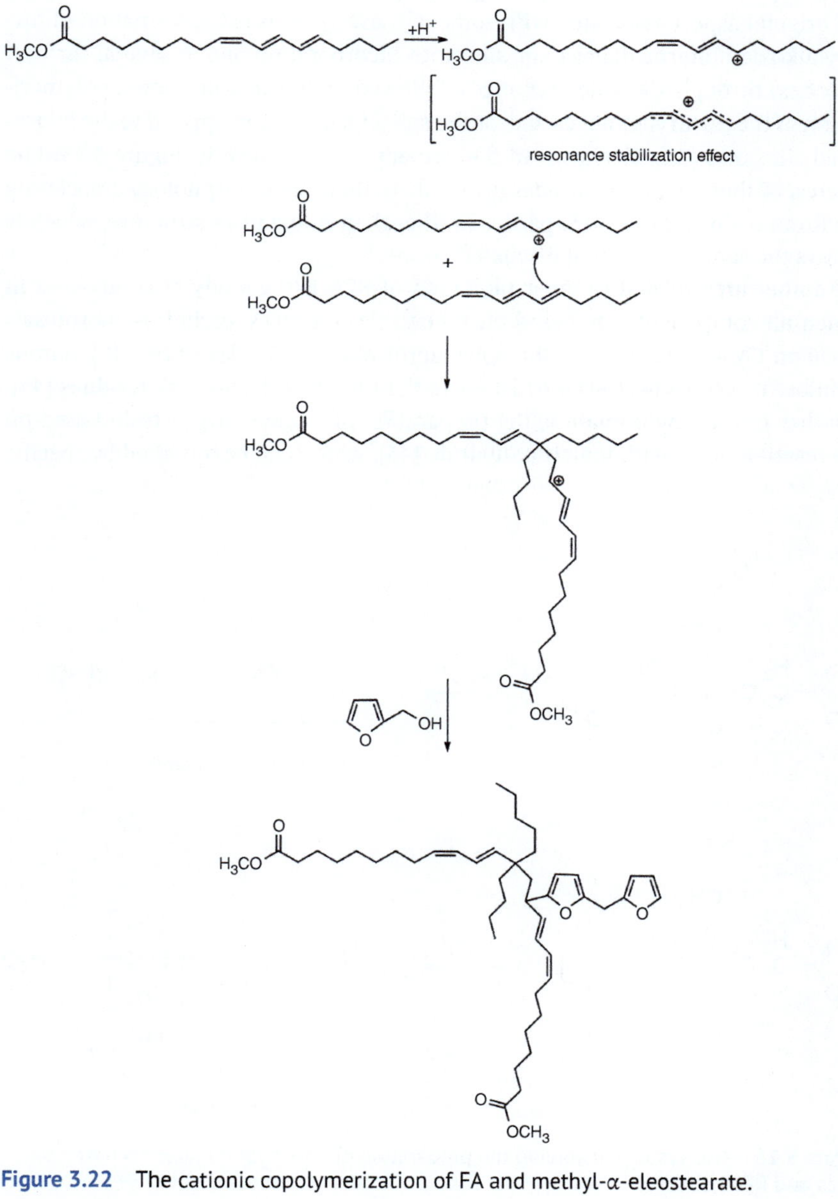

Figure 3.22 The cationic copolymerization of FA and methyl-α-eleostearate.

reactivity, favoring FA, the tung oil derivative was activated first and FA followed after the former monomer had already advanced in its polymerization.

A variety of other materials based on PFA have been prepared and characterized, including nanocomposites and hybrids of different compositions bearing controlled morphologies [34], including the synthesis of organic–inorganic hybrids [35], nanoscopic morphologies [36], and bio-based nanomaterials [37]. An original aspect associated with some of these systems is the formation of furfurylalkoxide intermediates (e.g., siloxanes incorporating one or several furfuryl moieties) through alkoxide exchanges, followed by the acid-promoted polymerization of the furfuryl moieties and/or the sol-gel mechanism applied to the hydrolyzed siloxanes [35], as shown in the two approaches given in Figure 3.23. The interest of these materials resides primarily in their nanomorphology associating the furan resin and the silica particle, rather than in the resin structure, which is always the same PFA as that discussed above.

Another area related to the exploitation of PFA is the study of composites in which all components are based on renewable resources, including thermosets based on FA and linseed oil [38], poly(caprolactone) [39], flax fibres [40], porous cellulose materials [41], sisal whiskers [42], humins [43], and cork residues [44]. Another interesting domain is the preparation of a nanoparticle resin based on the reaction of FA with maleic anhydride [45], which can be modified by specific

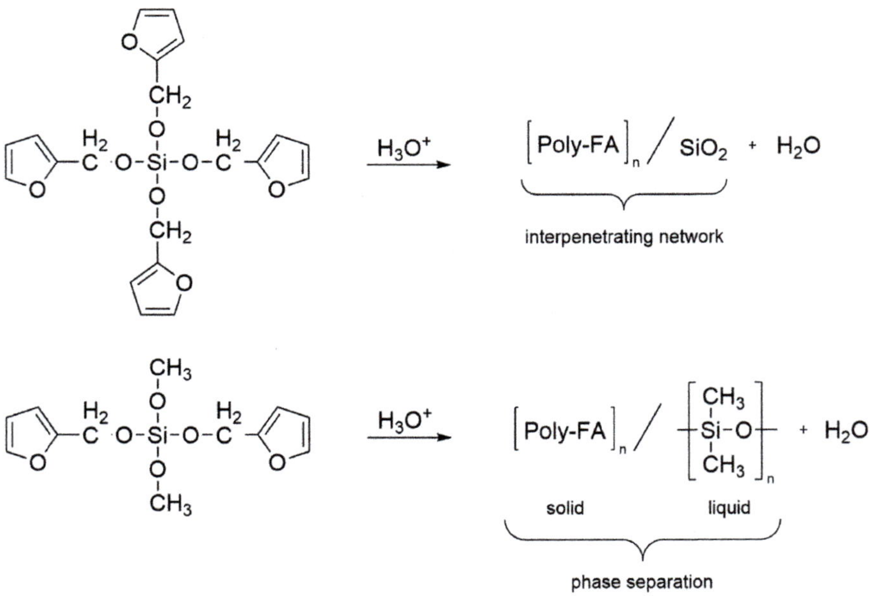

Figure 3.23 Two systems involving the preparation of nanohybrid materials based on silica and PFA networks.

surface grafting, e.g., with amines. The incorporation of bacterial cellulose into a PFA/tung oil co-network generated by the same cationic initiation as depicted in Figure 3.22, now applied to the corresponding triglyceride [46], provided a good dispersion of the nanofibers thanks to their introduction into the unreacted FA before the copolymerization. This procedure insured the good compatibility of the two monomers because it did not involve the use of the traditional water medium for the nanocellulose, which was replaced by FA through solvent exchange. Recently, gelatinized starch was mixed with the reaction product of FA with formaldehyde, which allowed it to react with the primary OH groups of the polysaccharide, and the ensuing material was finally mixed with microcrystalline cellulose to produce a fully biobased composite [47]. Although the chemical interactions involved in this system were not explored, it clearly appears that, whereas some of the FA was taken up by the coupling with starch, the rest polymerized in the acidic medium and its growing chains reacted with some the C5 positions of the starch-appended furan moieties, thus giving rise to a cross-linked material. Blowing this composite produced foams with good compression and char-generating properties, which were also shown to be biodegradable [47].

The preparation of PFA/cellulose-fibres composites has been successfully applied to 3D printable materials by extruding the prepolymer and optimizing the fibres content and the process parameters associated with the final cross-linking in order to attain a smooth and strong final print [48].

Yet another interesting exploitation of FA resins has been recently put forward [49], namely taking advantage of their structural imperfections, here the frequent units bearing open ring moieties with carbonyl groups, which were made to react with a fatty-acid based diamine to generate fully biosourced, highly cross-linked, heat resistant networks with tough and ductile properties.

The varied PFA field of useful applications reflects a healthy situation which promises future stimulating advances.

References

1 (a) Dunlop, A.P. and Peters, F.N. (1953). *The Furans. ACS Monograph Ser.* 119. Reinhold Publishing Co., New York. (b) McKillip, W.J. (1989). *ACS Sym. Ser.* 385: 408–423. (c) Hoydonckx, H.E., Van Rhijn, W.M., Van Rhijn, W., De Vos, D.E., and Jacobs, P.A. (2008). Furfural and Derivatives. In: *Ullmann's Encyclopedia of Industrial Engineering*, Vol. 15. Wiley-VCH, Weinheim, 1–29. (d) Gandini, A. and Belgacem, M.N. (2003). Furan-based Adhesives. In: *Handbook of Adhesive Technology*, 2e. (ed. A. Pizzi and K.L. Mittal). Marcel Dekker, New York, Ch. 30, 615–634. (e) Eseyin, A.E. and Steel, P.H. (2015). *Int. J. Adv. Chem.* 3: 42–47. (f) Gandini, A. and Belgacem, M.N. (2022). Furan Resins. In: *Handbook of Thermoset Plastics*, 4e. (ed. A. Dodiuk). Elsevier, Oxford (UK), 83–96.

2 Galego, N. and Gandini, A. (1975). *Revista CENIC* (Cuba) 6: 163–180.
3 (a) Sanchez, R., Hernandez, C., Jalsovsky, G., and Czira, G. (1994). *Eur. Polym. J.* 30: 37–42. (b) Sanchez, R., Hernandez, C., and Keresztury, G. (1994). *Eur. Polym. J.* 30: 43–50. (c) Sanchez, R. and Hernandez, C. (1994). *Eur. Polym. J.* 30: 51–54.
4 Galego, N. and Gandini, A. (1984). *Revista CENIC* (Cuba) 15: 143–146.
5 Rivero, G., Fasce, L.A., Ceré, S.M., and Manfredi, L.B. (2014). *Progr. Org. Coat* 77: 247–256.
6 (a) Rivero, G., Vazquez, A., and Manfredi, L.B. (2010). *J. Appl. Polym. Sci.* 117: 1667–1673. (b) Rivero, G., Pettarin, V., Vázquez, A., and Manfredi, L.B. (2011). *Thermochim. Acta* 516: 79–87.
7 Cayh, G. and Küsefoglu, S. (2011). *J. Appl. Polym. Sci.* 120: 1707–1712.
8 (a) Rodriguez, V.J. and Gandini, A. (1974). *Revista CENIC* (Cuba) 5: 29–39. (b) Koršak, V.V., Cejtlin, G.M., Chomutov, V.A., and Gonzales, C.P. (1980). *Acta Polym.* 31: 587–591.
9 Rodriguez, V.J. and Gandini, A. (1975). *Revista CENIC* (Cuba) 6: 155–161.
10 Salon, M.C., Gandini, A., and Cheradame, H. (1984). *Polym. Bull.* 12: 441–448.
11 Martinez-Garcia, A., Ortiz, M., Martinez, R., Ortiz, P., and Reguera, R. (2004). *Ind. Crops Prod.* 19: 99–106.
12 Sharifat, M., Efhamisisi, F., Thevenon, M.-F., Tarmian, A., and Jonoobi, M. (2021). *BioResources* 16: 7139–7160.
13 Ghafari, R., DoostHosseini, K., Abdulkhani, K.A., and Mirshokraie, S.A. (2016). *Eur. J. Wood Wood Prod.* 74: 609–616.
14 Gómez Millán, G. and Sixta, H. (2020). *Catalysts* 10: 1101. DOI: 10.3390/catal10101101.
15 Choura, M., Belgacem, N.M., and Gandini, A. (1996). *Macromolecules* 29: 3839–3850.
16 Kim, T., Assary, R.A., Marshall, C.L., et al. (2011). *ChemCatChem* 3: 1451–1458.
17 (a) Wewerka, E.M., Longhran, E.D., and Waters, K.L. (1971). *J. Appl. Polym. Sci.* 15: 1437–1451. (b) Wewerka, E.M. (1971). *J. Polym. Sci.* A19: 2703–2715. (c) Milkovic, J., Myers, G.E., and Young, R.A. (1979). *Cellulose Chem. Technol.* 13: 651–658.
18 (a) Buchwalter, S.L. (1987). *Polym. Prepr. (Am. Chem. Soc., Div. Polym. Chem.)* 28: 211–213. (b) Buchwalter, S.L. (1985). *J. Polym. Sci.* 23: 2897–2911.
19 (a) Conley, R.T. and Metil, I. (1963) *J. Appl. Polym. Sci.* 7: 32–52. (b) Maciel, G.E., Chuang, I.S., and Myers, G.E. (1984) *Macromolecules* 17: 1087–1090.
20 (a) Stagnaro, P., Costa, G., and Gandini, A. (2001). *Macromolecules* 34: 26–32. (b) Stagnaro, P., Costa, G., Pioli, F., and Gandini, A. (2003). *Polymer* 44: 1359–1365. (c) Stagnaro, P., Pioli, F., Panizza, M., and Gandini, A. (2009). *Macromolecules* 42: 2455–2461. (d) Stagnaro, P., Panizza, M., Gandini, A., Prevosto, D., and Lucchesi, M. (2011). *J. Polym. Sci., Polym. Chem.* 49: 5227–5238.
21 Schmitt, C.R. (1974). *Polym.-Plast. Technol.* 3: 121–158.
22 Lagel, M.C., Zhang, J., and Pizzi, A. (2015). *Ind. Crops Prod.* 67: 264–269.
23 Lagel, M.C., Hai, L., Pizzi, A., et al. (2016). *Ind. Crops Prod.* 85: 372–381.

24 Guigo, N., Mija, A., Vincent, L., and Sbirrazzuoli, N. (2010). *Europ. Polym. J.* 46: 1016–1023.

25 (a) Zhu, X., Bruijnaers, B., Lourençon, T.V., and Balakshin, M. (2022). *Materials* 15: 350. DOI: 10.3390/ma15010350. (b) Xu, J., Hu, D., Zheng, Q., Meng, Q., and Li, N. (2022). *Polymers* 14: 1071. DOI: 10.3390/polym14061071.

26 Luckeneder, P., Gavino, J., Kuchernig, R., Petutschnigg, R., and Tondi, G. (2016). *Polymers* 8: 396. DOI: 10.3390/polym8110396.

27 (a) Lande, S., Eikenes, M., and Westin, M. (2004). *Scand. J. For. Res.* 19: 14–21. (b) Lande, S., Westin, M., and Schneider, M. (2004). *Scand. J. For. Res.* 19: 22–30. (c) Lande, S., Westin, M., and Schneider, M. (2008). *Mol. Cryst. Liq. Cryst.* 484: 367–378.

28 (a) Li, W., Wang, H., Ren, D., Yu, Y.S., and Yu, Y. (2015). *Wood Sci. Technol.* 49: 845–856. (b) Zelinka, S.L., Altgen, M., Emmerich, L., et al. (2022). *Forests* 13: 1004. DOI: 10.3390/f13071004.

29 Martin, L.S., Jelavic, S., Cragg, S.M., and Thygesen, L.G. (2021). *Green Chem.* 23: 8003–80015.

30 Zheng, Z., Chen, C., Huang, Y., et al. (2022). *ACS Appl. Polym. Mater.* 4: 5308–5318.

31 Dong, Y., Wang, K., Li, J., Zhang, S., and Shi, S.Q. (2020). *ACS Sustainable Chem. Eng.* 8: 3532–3540.

32 Pizzi, A. and Khan, A. (2021). Furanic Rigid Foams, Furanic-Based Bioplastics and Furanic-Derived Wood Adhesives and Bioadhesives. In: *Furan Derivatives – Recent Advances and Applications*. (eds. A. Khan et al.). IntechOpen, London. DOI: 10.5772/intechopen.101200.

33 Ribeiro, B.O., Valério, V.S., Gandini, A., and Lacerda, T.M. (2020). *Int. J. Biol. Macromol.* 164: 2497–2511.

34 (a) Wang, H. and Yao, J. (2008). *Ind. Eng. Chem. Res.* 45: 6393–6404. (b) Casano, F., Scarano, S. Bertarione, S., et al. (2008). *J. Photochem. Photobiol. A Chem.* 196: 143–153. (c) Bertarione, S., Bonino, F., Cesano, F., et al. (2009). *J. Phys. Chem. B* 113: 10571–10574. (d) Yao, J. and Wang, H. (2007). *Ind. Eng. Chem. Res.* 46: 6264–6268. (e) He, L., Li, D., Zhang, G., et al. (2010). *Ind. Eng. Chem. Res.* 49: 4175–4180.

35 (a) Zhai, Y., Tu, B., and Zhao, D. (2009). *J. Mater. Chem.* 19: 131–140. (b) Edwards, E.R., Oishi, S.S., and Botelho, E.C. (2018). *Polimeros* (Brazil) 28: 15–22.

36 (a) Grund, S., Kempe, P., Baumann, G., Seifert, A., and Spange, S. (2007). *Angew. Chem. Int. Ed.* 46: 628–632. (b) Spange, S. and Grund, S. (2009). *Adv. Mater.* 21: 2111–2116. (c) Uhlig, D., Seifert, A., Schreiter, K., et al. (2019). *Macromol. Chem. Phys.* 220: 1900050.

37 (a) Pranger, L.A. and Tannenbaum, R. (2008). *Macromolecules* 41: 8682–8687. (b) Pranger, L.A., Nunnery, G.A., and Tannenbaum, R. (2012). *Composites B* 43: 1139–1146.

38 Pin, J.-M., Guigo, N., Vincent, L., Siirrazzuoli, N., and Mija, A. (2015). *ChemSusChem* 8: 4098.

39 (a) Sommerauer, L., Grzybek, J., Elsaesser, M.S., et al. (2019). *Polymers* 11: 1533. (b) Nanni, G., Heredia-Guerrero, J.A., Paul, U.C., et al. (2019). *Polymers* 11: 1069.

40 (a) Pohl, T., Bierer, M., Natter, E., et al. (2011). *Plast. Rubber Compos.* 40: 294–299. (b) Deka, H., Misra, M., and Mohanty, A. (2013). *Ind. Crops Prod.* 41: 94–101.

41 Lems, E.-M., Winklehner, S., Hansmann, C., Gindl-Altmutter, W., and Veigel, S.R. (2019). *Cellulose* 26: 4431–4444.

42 Ahmad, E.E.M., Luyt, A.S., and Djoković, V. (2013). *Polym. Bull.* 70: 1265–1276.

43 Pin, J.-M., Guigo, N., Mija, A., et al. (2014). *ACS Sustain. Chem. Eng.* 2: 2182–2190.

44 Menager, C., Guigo, N., Wu, X., Vincent, L., and Sbirrazzuoli, N. (2019). *Compos. Part Appl. Sci. Manuf.* 124: 105473.

45 Luo, W., Liu, J., Ma, Y., Zhang, B., and Yang, W. (2012). *J. Polym. Sci., A, Polym. Chem.* 50: 3306–3317.

46 Valentino, H.A.S., Pupio, P.T.L.R.S., Gandini, A., and Lacerda, T.M. (2021). *Cellulose* 28: 7109–7121.

47 Liu, B., Essawy, H., Deng, S., et al. (2023). *Ind. Crops Prod.* 194: 116282. DOI: 10.1016/j.indcrop.2023.116282.

48 (a) Beneventi, D., Chaussy, D., Bouzidi, K., and Gandini, A. (2021). Patent *WO2021151893 (A1) – Thermosetting Composite Material and Associated Printing Process,* to Grenoble Polytechnic, Grenoble, France. (b) Bouzidi, K., Chaussy, D., Gandini, A., Bongiovanni, R., and Beneventi, D. (2022). *Carbohydr. Polym.* 293: 119716. DOI: 10.1016/j.carbpol.2022.119716. (c) Bouzidi, K., Beneventi, D., Flahaut, E., et al. (2022). *Compos. Sci. Technol.* 230: 109765. (d) Bouzidi, K., Beneventi, D., Gandini, A., Bongiovanni, R., and Chaussy, D. (2023). *J. Polym. Sci.*, 13 May 2023 https://doi.org/10.1002/pol.20230074.

49 Dellière, P. and Guigo, N. (2022). *ACS Macro Lett.* 11: 1202–1206.

4

Polymers from Chain Polyaddition Reactions

This chapter deals with the use of such systems as free radical, cationic, anionic, and stereospecific polymerizations and copolymerizations involving most of the monomers in Figure 2.19 and of some others like furan itself, and with the structure and properties of the ensuing macromolecular materials.

4.1 Free Radical Systems

Whereas furan does not polymerize via radical initiation, because of the reactions shown in Figures 2.3 and 2.4, giving at most some dihydrofuran oligomers, the furan-maleic anhydride DA adduct shown in Figure 2.14 behaves like a monomer in the presence of free radical initiators giving two types of macromolecular structures, depending on the reaction temperature. Above 70 °C, the polymer has the structure shown in Figure 4.1 resulting from the activation of the charge-transfer complex, associated with the DA adduct, displayed in Figure 4.2 [1]. At temperatures below the thermal decomposition of the adduct, the ensuing fully-saturated polymer structure reflects the fact that this is the actual monomer, as shown in Figure 4.3 [1]. It is noted that 2,5-dimethylfuran behaves similarly [1]. The DPs of these materials are modest, comprised between 10 and 30 and, to the best of our knowledge, no further study has dealt with their properties and possible applications.

Other copolymers of furan and alkylfurans, prepared by free-radical initiation have been investigated, mostly with acrylonitrile [2], but yields and DPs were low and the modest incorporation of the furan monomer into the polymer chains included various structural forms. With the strongly electron-accepting methyl-α-cyanoacrylate [3], different feed ratios always gave copolymers of regular structure, which contained the near-alternating moiety shown in Figure 4.4 when the furan concentration was increased, but at the expense of the yield.

Furan Polymers and their Reactions, First Edition. Alessandro Gandini and Talita Martins Lacerda.

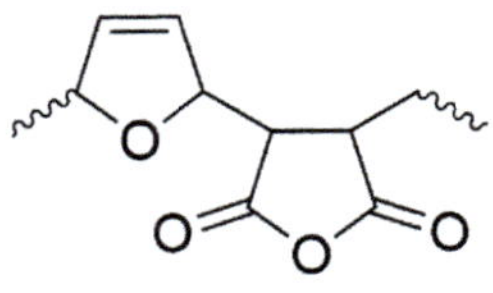

Figure 4.1 The structure of the furan-maleimide polymer formed at high temperature.

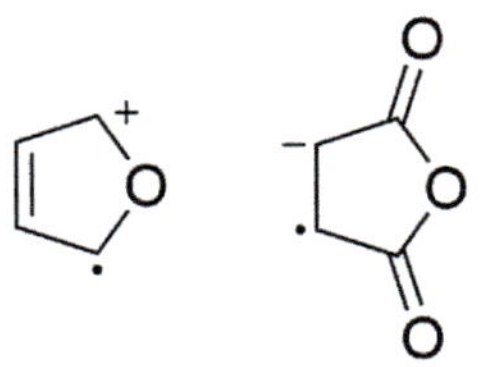

Figure 4.2 The charge-transfer complex associated with the furan-maleimide DA adduct.

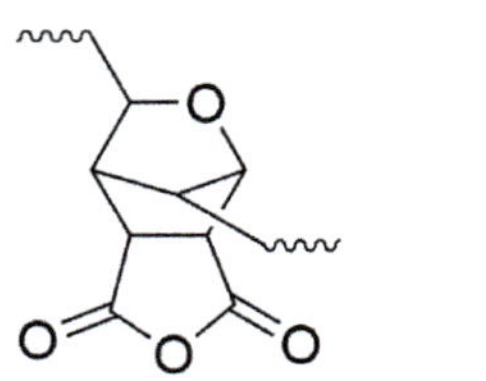

Figure 4.3 The structure of the polymer formed by the furan-maleimide adduct below its thermal decomposition.

On the whole, however, the retarding role of the heterocycle made these systems sluggish in their performance, both in terms of polymer yields and molecular weights.

The situation with the furan monomers in Figure 2.19, which can in principle undergo polyaddition reactions giving the generic macromolecular structure given in Figure 2.20, is very strongly determined by the possible competition that a free radical is faced between reacting with the alkenyl moiety external to the ring and the heterocycle itself, given the dienic character of the latter, which gives rise to the formation of the stabilized furyl radicals shown in Figure 2.3 [4]. This sensitive context was the object of a systematic study which was carried out using two approaches, namely (i) the role of various simple furans to conventional free-radical systems and (ii) the response of different alkenyl furans to free-radical activation, as a function of their specific structure [5]. Concerning the first approach, Table 4.1 summarizes the typical response of three classical monomers undergoing conventional free-radical polymerization conditions in the presence of 2-substituted furans. Kinetic data, as well as the use of a wider range of both monomers and furan additives, including the already mentioned investigation of the reaction of AIBN with furan compounds [4], confirmed more quantitatively these general features, which clearly indicate that the retarding or inhibiting role of furans in otherwise normal free-radical polymerizations critically depends on the structure of the involved monomer. Thus, those monomers which generate well-stabilized radicals are insensitive to the presence of even high proportions of the added furan compound, whereas, at the other extreme, those monomers characterized by poorly stabilized growing radicals witness an inhibiting effect to their polymerization when even modest amounts of furan additives are present. Here, when acrylic or methacrylic monomers are involved, associated with stabilized radicals, both initiation and propagation steps are not perturbed by the presence of the added furan, while vinyl acetate and vinyl chloride, known to generate weakly stabilized radicals, witness a strong inhibition of their polymerization induced by the competing interference of the added furan, which dominates as radical captor. Interestingly, in the case of styrene the situation is intermediate, which suggests that the degree of stabilization of the styryl and furyl radicals is similar, thus giving retardation or inhibition, depending on the concentration of the perturbing furan compound.

These conclusions apply to both initiation and propagation reactions, as sketched in Figure 4.5.

In the second set of studies [5], furan alkenyl monomers were submitted to typical free-radical initiation conditions and their quantitative response was assessed. 2-furfuryl methacrylate gave a "normal" polymerization, entirely equivalent to that of methyl methacrylate, whereas 2-vinyl furoate simply did not polymerize. Again, 2-vinyl furan displayed an intermediate behavior by showing a "normal" initial polymerization, which however stopped at limited conversion by a self-retarding effect, suggesting that the vinyl moiety competes effectively with the furan ring for free radical addition, but that the furan moieties of the accumulating polymer progressively become a trap for the radicals and hence impede any further 2-vinyl furan polymerization.

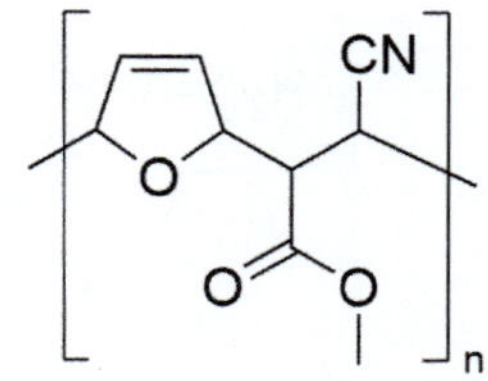

Figure 4.4 The alternating furan/methyl-α-cyanoacrylate copolymer prepared by AIBN photochemical initiation at room temperature.

In this context the primary radicals encounter two options for attack, namely, the alkenyl moiety or the C5 position of the heterocycle, as shown in Figure 4.6, and the issue, viz. initiation or inhibition, is determined by the relative stability of the ensuing radicals. Of course, these considerations apply equally to the alternative modes of propagation.

The results of copolymerization systems [5] were consistent with all the above evidence, displaying a straightforward behavior in the case of 2-furfuryl methacrylate with methyl methacrylate or styrene, an inhibiting or retarding role of 2-vinyl furoate with styrene, and reduced yields of methyl methacrylate homopolymer when the furan monomer was present in high feed ratios. With 2-vinyl furan (M_1), its copolymerization with styrene and acrylonitrile (M_2) gave $r_1 = 1.9$, $r_2 = 0.25$ and $r_1 = 0.82$, $r_2 = 0.04$, respectively, indicating a stronger tendency for M_1

Table 4.1 The effect of furan additives on the free-radical polymerization of three conventional monomers.

Monomer	*Furan compound (%, v/v)*	*Result*
Methyl acrylate	Ethyl furoate (25)	No effect
Methyl acrylate	2-Methylfuran (50)	No effect
Styrene	Ethyl furoate (20)	Retardation
Styrene	Ethyl furoate (50)	Inhibition
Vinyl acetate	Ethyl furoate (20)	Inhibition
Vinyl acetate	2-Methylfuran	Inhibition

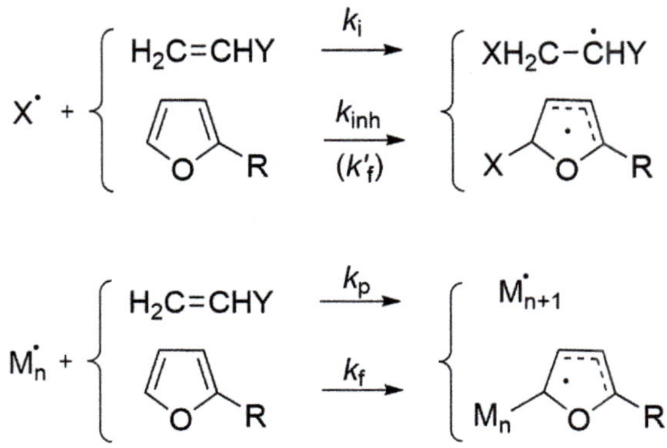

Figure 4.5 The competition of normal initiation and propagation in the free-radical polymerization of alkenyl monomer by the perturbing role of an added furan compound.

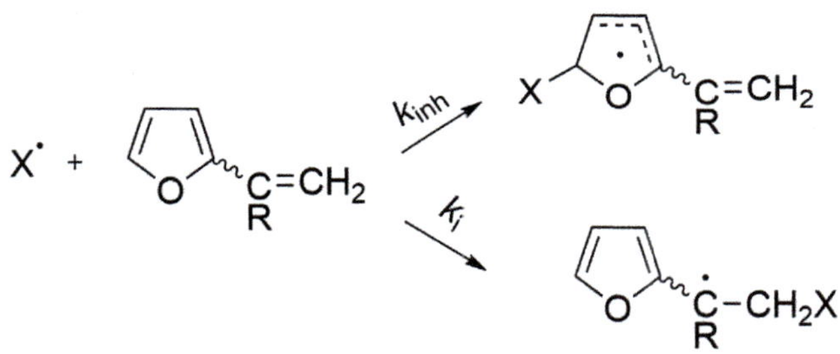

Figure 4.6 Two alternative modes of reaction of a primary radical with a furan-alkenyl monomer.

incorporation, i.e., 2-vinyl furan homopolymerization, but yields were low due to the progressive inhibiting role of the M_1 units in the copolymer, as already witnessed in its polymerization.

Interestingly, the emulsion polymerization of 2-vinylfuran proceeds to high yields and molecular weights, in contrast with the more conventional solution and bulk processes [6]. The reasons for this difference in behavior is probably to be ascribed to the fact that in the emulsion system, only one free radical (or none) is present in each active micelle.

This general state of affairs gives rise to two major contrasting conclusions. On the one hand, only furan alkenyl monomers giving rise to a dominant initiation through the external unsaturation will produce a conventional free-radical polymerization, as in the case of the acrylic structures, further discussed below. On the other hand, when the heterocycle is preferentially attacked by the primary radicals, polymerization is strongly retarded or inhibited, as in the case of 2-vinyl furoate. The intermediate behavior of 2-vinyl furan is interesting, but the corresponding

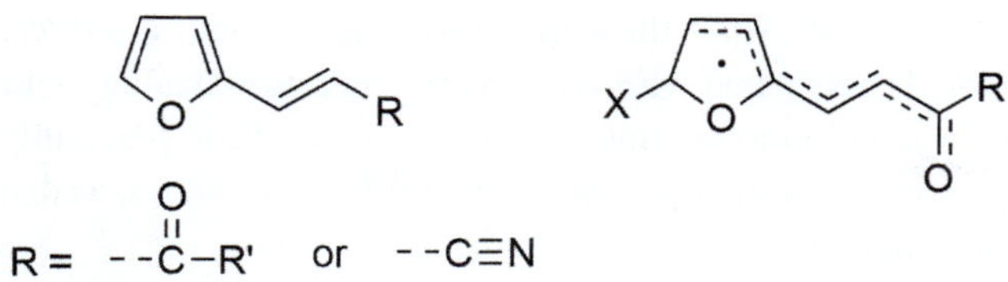

Figure 4.7 Examples of conjugated furan molecules as powerful free-radical traps thanks to the formation of highly stabilized radicals.

isolated polymer is prone to slow crosslinking because of the ease of the loss of the tertiary H-atom, which leaves a stabilized macroradical capable to couple with another one. Interestingly, its ATRP copolymerization with styrene [7] gave similar reactivity ratios and hence the possibility of synthesizing near-ideal copolymers, which were found to be stable for years because of the absence of long 2-vinyl furan monomer sequences.

A positive aspect which allows taking advantage of the inhibiting features of some of these structures is their extension to maximize them in view of preparing very powerful free-radical traps, by the use of cheap conjugated furan molecules like those shown in Figure 4.7, which were tested in systems involving free-radical intermediates, such as polymerization, oxidation, and photolysis, also including polymer degradation processes [8].

The free-radical polymerization and copolymerization of furfuryl acrylate and methacrylate have been extensively studied [9] and revealed systematically that they proceed just as those of the other well-characterized members of this large family of monomers, i.e., that the presence of the furan ring does not interfere with the regular polymer growth. As hinted above, this stems from the stabilization of the active radical by the ester function, just like with conventional acrylates and methacrylates, giving the structure shown in Figure 4.8 without detectable interference from possible reactions with the heterocycle. It follows that high yields and molecular weights are obtained and only some minor crosslinking is observed at very high conversions [9]. Thus, for example, furfuryl methacrylate typically gives a polymer with a Mn of some 90,000 Da and a Tg of 50 °C, while its TGA in an inert atmosphere revealed the onset of degradation at 250 °C with both depolymerization and intramolecular cyclization releasing FA. In air, oxidative crosslinking also takes place. Its copolymerization with 2-hydroxyethyl methacrylate (M2) gave $r1 = 0.9$ and $r2 = 1.3$ and the expected regular variation of Tg between 50 and 90 °C for the M2 homopolymer. Hydrogels prepared from these copolymers were assessed as potential materials

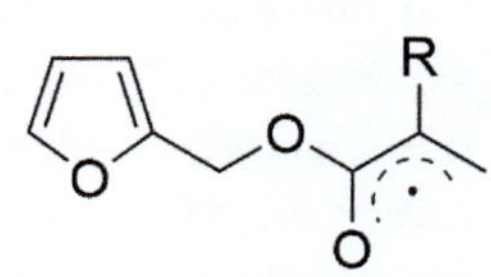

Figure 4.8 The stabilized structure of the furanacrylic monomer radical. R = H or CH_3.

for biocompatible applications [9]. Since these investigations, a renewed interest in furanacrylic monomers is associated with the growing literature dealing with the application of the DA reaction to polymer science because of the possibility of readily constructing copolymers incorporating pendant furan moieties, as discussed in the appropriate chapter.

Two other furan monomers are worth a brief mention concerning their response to free-radical initiation, namely 2-furyl vinyl ketone [10] and 2-furyl oxirane [11], which however responded poorly to this type of activation, giving low yields of oligomers. On the other hand, 2-vinyl-5-hydroxymethyl furan and its 5-methoxy homologue have been successfully polymerized by AIBN in bulk at 70 °C [12] to give regular low-Mn materials with poor thermal stability and no mention was provided of their atmospheric shelf life in terms of structural evolution. These same monomers were reinvestigated in a different context [13] in the sense that the 5-hydroxy homologue was found to display exceptional adhesive properties when applied as monomer and heated at 110 °C for two hours, during which time the authors postulated that it had polymerized via a free-radical mechanism. The binding power, similar to that of cyanoacrylates, and more efficient than that of commercial epoxy resins, was attributed to the -OH groups borne by each monomer unit, because the 5-methoxy homologue did not exhibit this property when subjected to the same curing process, but also to crosslinking arising from some monomer dimerization through the -CH_2OH group that produced divinyl molecules.

The lack of polymerizability of 2-vinyl furoate is not in itself a barrier to the synthesis of the corresponding polymer, which can be readily prepared by the esterification of poly(vinyl alcohol) with furoyl chloride, albeit not at 100% yield [14].

4.2 Cationic Systems

Furan and alkylfurans were the first potential monomers investigated in the context of cationic activation in the 1960s using both Brønsted and Lewis acids [5]. The behavior of 2-methylfuran was at first interpreted in terms of a simple propagation through one of the ring unsaturation giving a linear dihydrofuran polymer, but this was later shown to be incorrect thanks to a more detailed structural assessment [15], which revealed the presence of both furan and dihydrofuran units, as well as evidence for ring-opening reactions giving -OH groups in a mixture of oligomers with Mn varying between 300 and 1300 Da. These authors also isolated the furan-aliphatic tetramer shown in Figure 4.9.

The cationic oligomerization of furan received a closer attention [16] using a variety of acids and media, including water [17, 18]. In strong acidic conditions the process gives colored products which slowly crosslink. The initiation step gives the cationic active species in Figure 4.10, which then propagates to give essentially the growing dihydrofuran oligomers depicted in Figure 4.11.

As the process advances, two further reactions are shown to intervene, namely (i) the intervention of the remaining unsaturations in the dihydrofuran units in the propagation, albeit at a much slower pace, producing branched structures and eventually crosslinking (Figure 4.12), and (ii) the opening of some of the dihydrofuran rings, giving rise to conjugated moieties, which would explain the progressive and deepening of color in the products (Figure 4.13). When the reactions were stopped before the sensible advance of these side reactions, the isolated oligomers had a DPn of about 10 and a DPw of about 50, reflecting however the importance of branching, even at this stage. Interestingly, furan groups were detected spectroscopically and were attributed to end group moieties.

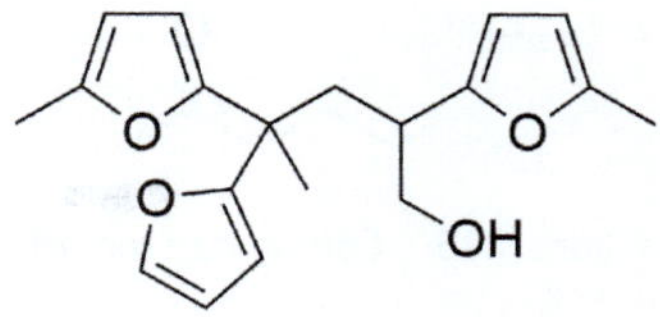

Figure 4.9 Tetramer isolated in the cationic polymerization of 2-methylfuran.

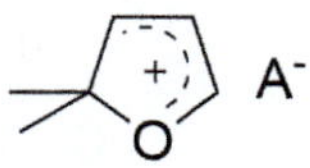

Figure 4.10 The protonated furan responsible for the initiation of its cationic oligomerization.

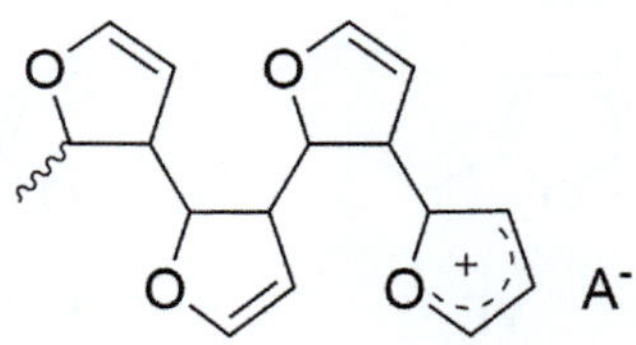

Figure 4.11 The growing dihydrofuran species in the furan cationic polymerization.

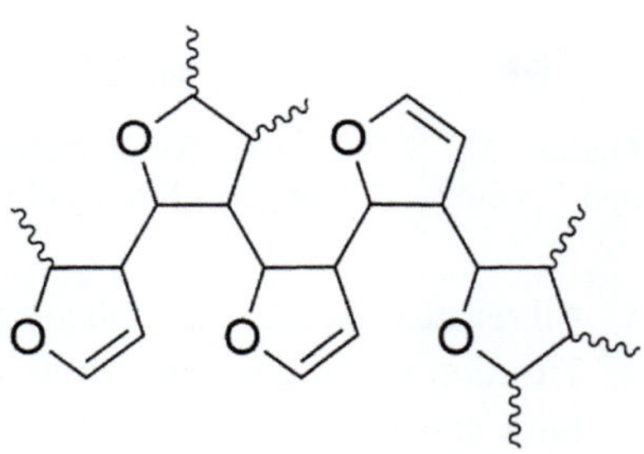

Figure 4.12 Branched structures in advanced furan cationic polymerization.

In aqueous HCl, furan was found to produce two major products consisting of the tetramers shown in Figure 4.14 [17, 18], arising from the role of hydronium ions in inducing the opening of a furan heterocycle.

2-Alkenylfurans constitute an interesting set of monomers because of their peculiar behavior in cationic polymerization, which was systematically studied in terms of their mechanisms, kinetics, and polymer structures and properties. Four monomers (Figure 4.15) were synthesized and fully characterized [19] before tackling a thorough investigation of their polymerization under vacuum, initiated mostly by trifluoroacetic acid, but also by other Brønsted and Lewis acids, mostly in methylene chloride, but also in solvents of lower and higher polarity, at temperatures comprised between −78 and 40 °C [20].

The following specific features describe the essential qualitative behavior of VF polymerizations:

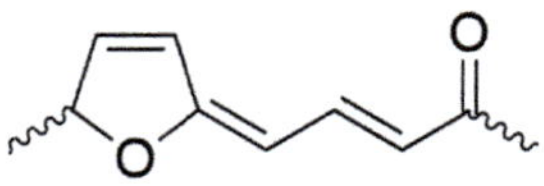

Figure 4.13 Conjugated moieties in advanced furan cationic polymerization.

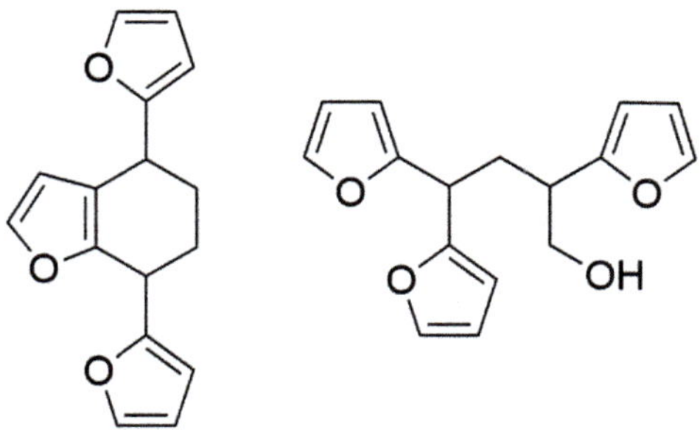

Figure 4.14 The two tetramers formed in the furan cationic polymerization in aqueous HCl.

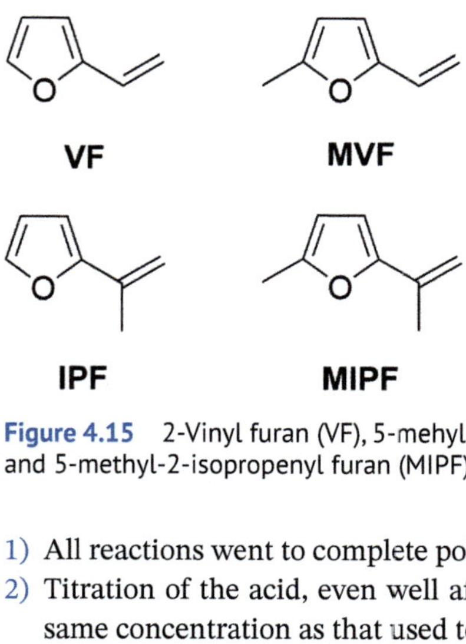

Figure 4.15 2-Vinyl furan (VF), 5-mehyl-2-vinyl furan (MVF), 2-isopropenyl furan (IPF) and 5-methyl-2-isopropenyl furan (MIPF).

1) All reactions went to complete polymer conversion.
2) Titration of the acid, even well after the end of the polymerization, gave the same concentration as that used to start the reaction.
3) Treatment of the reaction medium with solid potassium hydrogen carbonate produced the total neutralization of the acid.
4) Addition of a second amount of monomer at the end of a given polymerization resulted in further polymerization, albeit at a lower rate.
5) Polymerization media turned progressively from pale yellow at low conversion to an amber color at later stages, then to dark brown towards the end of the reaction. The color continued to deepen even after complete conversion, reaching dark blue-grey tonalities, and even blackish hues if the solutions were kept for weeks before neutralization.

6) The electrical conductivity of the polymerizing solutions grew steadily as the reaction advanced and kept growing even after complete consumption of the monomer.
7) Neutralization of the deeply-colored conducting solutions induced a rapid partial bleaching and a drastic drop in conductivity, leaving a light orange-brown, poorly conducting medium. Re-acidification brought back the original color and high conductivity. This cycle could be repeated indefinitely.
8) If an empty side arm of the reaction vessel was cooled down during the polymerization, a fraction of the three volatile component of the reaction mixture vacuum-distilled into it as shown by the onset of further polymerization, albeit at a slower pace, accompanied by the same color and conductivity features.
9) The yellow-brown polymer isolated from these systems were in fact oligomers with DPn ranging from 5 to 20, depending on the reaction conditions, which displayed important structural irregularities with respect to the expected regular poly(2-vinyl furan) of Figure 4.16.

The polymerization of the other three monomers displayed both common features regarding complete conversion, lack of acid consumption and re-initiation of the process after a second monomer addition, and very different ones regarding color and conductivity, but mostly in terms of polymer structure, as discussed in detail below. In fact, the decision to extend this study to MVF, IPF, and MIPF was dictated by the urge to understand the apparent anomalies in the behavior of VF, such as color development, increase in electrical conductivity and, most of all, the irregularities in its polymer structure.

The first of such irregularities has to do with the intervention of C5 electrophilic substitutions when that position is free (see Figure 2.5) [21]. Based on ^{1}H NMR spectroscopy of both VF polymerizing solution and the isolated polymer, accompanied by the synthesis and behavior of model compounds, this reaction involves the attack of a growing species on either the monomer or a polymer unit, as shown in Figure 4.17. In the first case, a furan ring is incorporated into the polymer chain, while giving an irregular isopropyl moiety, whereas in the second instance the reaction generates a branched polymer. As expected, the polymerization of IPF was accompanied by the same reactions, corroborated by spectroscopic evidence [21]. Conversely, MVF and MIPF produced linear polymers, confirming the high C5 regiospecificity of these reactions, since no C3 or C4 substitution was detected in any of these four polymers.

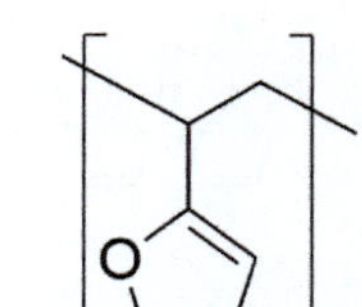

Figure 4.16 The expected normal structure of poly(2-vinyl furan).

The second irregularity relates to color formation accompanied by increasing ionic conductivity. A thorough investigation of the evolution of the UV-visible spectra of the polymerizing and neutralized solutions, together with ^{1}H NMR spectroscopy [22], revealed the occurrence of a reaction sequence involving

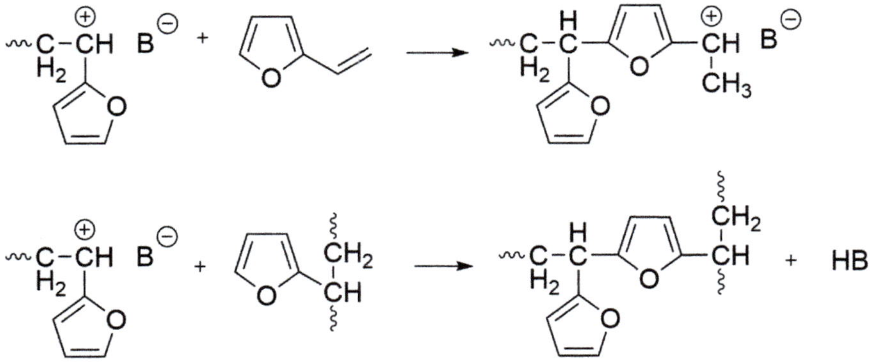

Figure 4.17 Two modes of C5 electrophilic substitution in the VF cationic polymerization. Here B^- represents the anion derived from the initiator.

the tertiary hydrogen atom in the aliphatic moiety of the growing polymer, which protonates an unsaturated end group of a dead polymer chain, as shown in Figure 4.18, giving a delocalized carbocation. This species is in equilibrium with the corresponding neutral conjugated unsaturation by loss of a proton. This sequence can then be repeated, thus increasing the extent of conjugation in both ionic and neutral form illustrated by Figure 4.18. These cycles explain the color formation and its bathochromic shift, as well as the ionic conductivity, which increases with

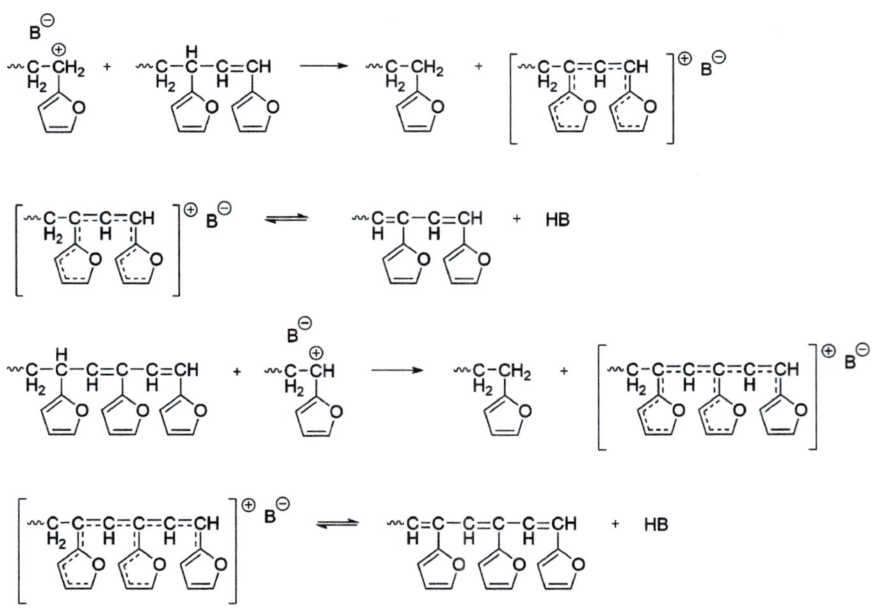

Figure 4.18 The sequence of events leading to conjugated end-structures in ionic and neutral form in the VF cationic polymerization.

the number of such events and with the increasing character of the free ions produced. It is important to emphasize that these reactions do not require the presence of the monomer and closely resemble the features described above in the case of the acid-catalyzed polycondensation of FA (Figure 3.17). As expected, MVF displayed the same features, whereas IPF and MIPF polymerized without producing color or conductivity increase, since in these monomers the tertiary hydrogen atom is replaced by a methyl group. In conclusion, only MIPF polymerized without appreciable secondary reaction to give the expected regular structure shown in Figure 4.19. In other words, by replacing the reacting H5 and Hα with methyl groups, the two side reactions could be entirely suppressed, proving irrefutably the causes of their occurrence.

The kinetics of these polymerizations and the nature of the active species were also assessed as a function of temperature and solvent polarity [23]. As expected, the monomer-transfer reactions, typical of the cationic polymerization of alkenyl monomers, could be considerably reduced, with respect to propagation, as the reaction temperature was lowered, thus enhancing the DP of the polymers.

A number of original strategies have been put forward, based on either of the two side reactions discussed here, which provided some positive remediation to their negative structural consequences in the polymerization of VF, MVF, and IPF. The C5 electrophilic substitution can be exploited as a source of dominant transfer reaction to prepare oligomers with furan end groups. Thus, for example, the cationic polymerization of isobutene was carried out in the presence of different concentrations of 2-methyl furan to produce oligomers of controlled DPn (from 2 to 100), as shown in Figure 4.20 [24]. The interest of this strategy is not limited to preparing well-defined oligomers for specific applications, such as temperature-independent viscosity materials (as in the case of isobutene), but also to exploit the furan end group as a source of further modification, e.g., through the DA reaction with maleic anhydride, or oxidation with RuO_4, that introduced polar and-groups on a non-polar polymer [24].

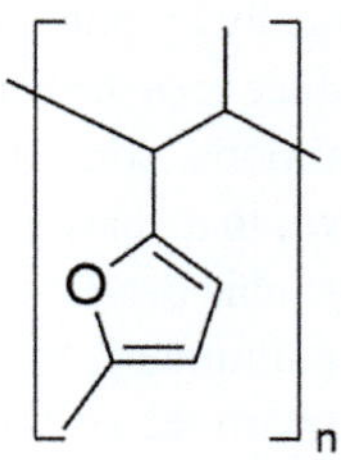

Figure 4.19 The regular structure of polyIPVF prepared by cationic polymerization.

In another approach, block copolymers can be prepared by a double C5 electrophilic substitution using difuran bridging agents, as sketched in Figure 4.21 [25],

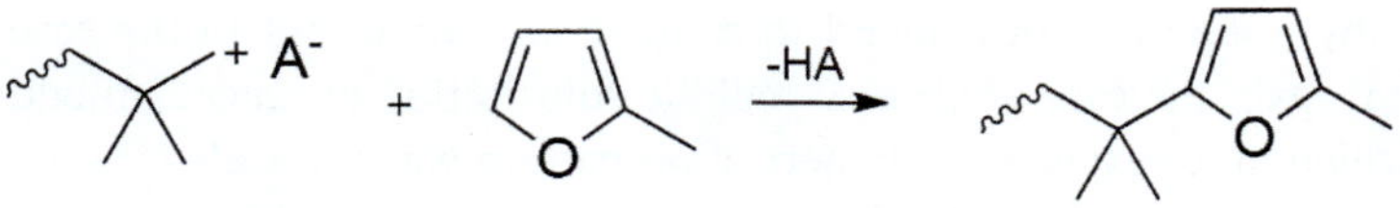

Figure 4.20 Synthesis of furan-terminated oligoisobutene by predominant chain transfer.

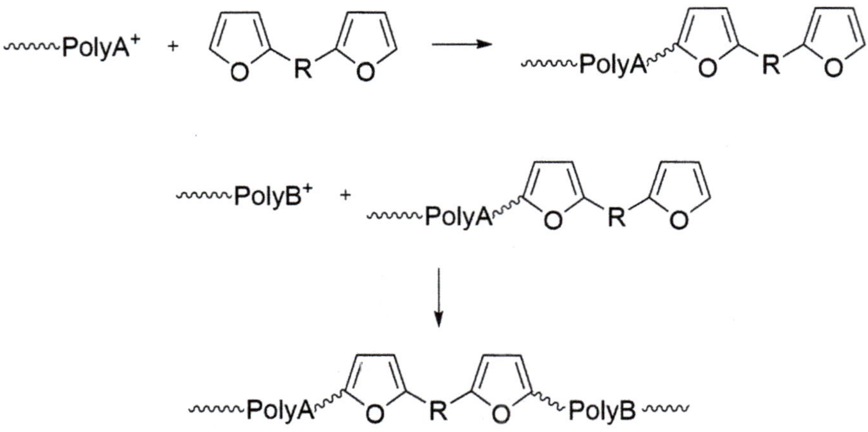

Figure 4.21 Synthesis of a block copolymer by double electrophilic substitution onto a difuran bridging compound.

where a first cationic polymerization is carried out with monomer A, followed by a second one involving monomer B.

Finally, within this context, grafting-onto reactions can be applied to polymers or copolymers bearing pendant furan rings by carrying out the cationic polymerization of a given monomer in their presence [24], as shown schematically in Figure 4.22.

The formation of highly conjugated and branched PolyVF can be enhanced dramatically by polymerizing VF in a very strong acidic medium, which can in fact produce crosslinked structures as shown in Figure 4.23, because in these drastic conditions, some C3 and C4 alkylation also take place. These materials were shown to display a very high proton affinity of up to 60 meq g^{-1} and can be used as recyclable proton exchange resins, as such [25], or coated on silica particles [26].

In summary, the thorough study and understanding of the intricacies associated with the cationic polymerization of 2-alkenyl furans highlighted once again the power of the dienic character of the heterocycle both in interfering with a normal growth behavior, but also in providing with novel ways of arriving at original macromolecular materials [24–26]. We believe that latter aspects deserve further attention, after a couple of decades of neglect.

The lack of response of 2-vinyl furoate to free-radical polymerization, discussed in the previous section, excluded it as a viable monomer in that context. It is well known that vinyl esters do not respond to cationic activation, but in the case 2-vinyl furoate, again the ease of C5 electrophilic substitution produced an odd situation in which an unexpected polymerization pattern was unraveled. When this monomer was submitted to typical cationic polymerization conditions, using $ZnCl_2$ or $BF_3.Et_2O$ in methylene chloride at room temperature, the protonation of the vinyl moiety of one of its molecule (initiation) created the conditions for the

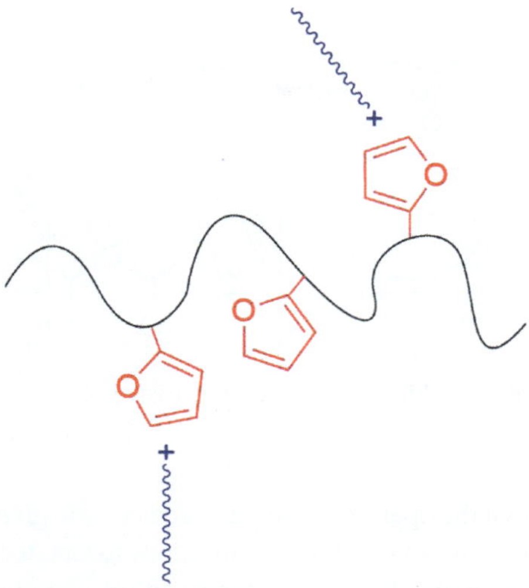

Figure 4.22 Grafted copolymer prepared by electrophilic C5 substitution.

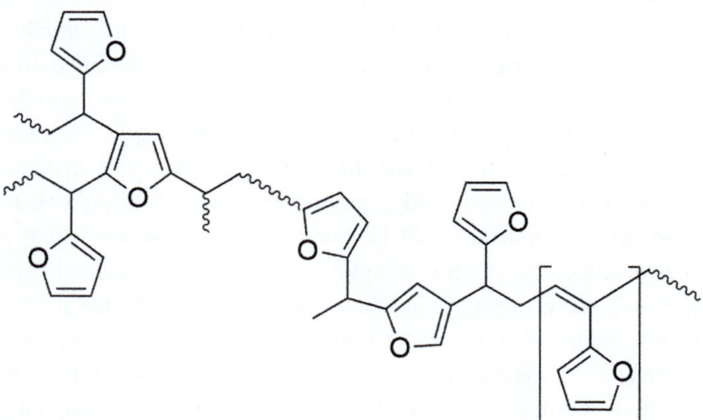

Figure 4.23 Crosslinked polyVF resins with high nucleophilicity.

subsequent C5 substitution of a second molecule (unconventional dimerization) which generated the same end-protonated species, thus promoting the pursuit of propagation to give a novel polyester with the furan ring incorporated into the polymer chain, as shown in Figure 4.24 [5a, 5b]. Interestingly, the structure of this polymer bears very little resemblance to the monomer that generated it, and it is also an exception to the correlation discussed above (Figure 2.20) in that the furan ring is here part of the polymer backbone. These results represent yet another example of the peculiar behavior of the furan heterocycle, which, regrettably, was not followed up despite the obvious interest of the system.

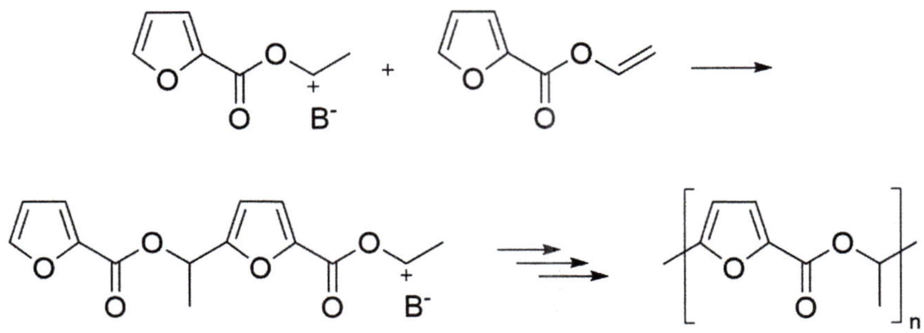

Figure 4.24 The cationic polymerization of 2-vinyl furoate through successive C5 electrophilic substitutions.

The polymerization of aldehydes through the opening of the carbonyl bond to give the corresponding poly(acetal)s is subject to thermodynamic restrictions associated with the low enthalpy of propagation and hence low ceiling temperatures, except for lower aliphatic aldehydes. Aromatic and highly conjugated aldehydes do not homopolymerize because of their very low ceiling temperatures, e.g., about −160 °C for benzaldehyde. Furfural (F) belongs to this family, and the various rather qualitative attempts to prepare its acetal homopolymer at low temperature have failed. In a more systematic study [27], F was submitted to a variety of conditions relative to the type of initiation, solvent, and temperature (down to −196 °C) and again the monomer was always recovered completely. A way to reduce this hindrance involves copolymerizing the unreactive aldehyde with a monomer possessing a higher ceiling temperature, and this was indeed tested [27] with cationic systems, where F was coupled with cyclopentadiene, styrene, α-methyl styrene, indene, acenaphthylene, butyl-vinyl ether, or *N*-vinyl carbazole in toluene or methylene chloride at 0 °C, using $BF_3.Et_2O$ as an initiator. With the lesser nucleophilic monomers, the F:BF_3.Et_2O complex inhibited their polymerization, whereas the highly nucleophilic ones produced exclusively their homopolymers. Only α-methyl styrene gave interesting copolymers in terms of F unit contents, which had values ranging from 30 to just above 50%. A systematic characterization of the 50/50 product clearly showed that it was an alternating copolymer with an Mn of 2000 Da. When F was replaced by MF, similar results were obtained, except that its copolymerization with α-methyl styrene (MST) gave products which could contain between 50 and 74% of MF units, suggesting that a penultimate effect operated in these systems allowing the incorporation of two MF repeated units [28]. Depending on the feed ratio, therefore, the structures of these polymers went from an alternating pattern to dominating MST-MF-MF sequences, as shown in Figure 4.25. Attempts to prepare copolymers between F or MF and aliphatic aldehydes failed.

In a second study [28], MF, which is more nucleophilic than F, was copolymerized with various vinyl ethers in the same conditions, and again copolymers with average Mn of a few thousand were obtained with the essentially alternating

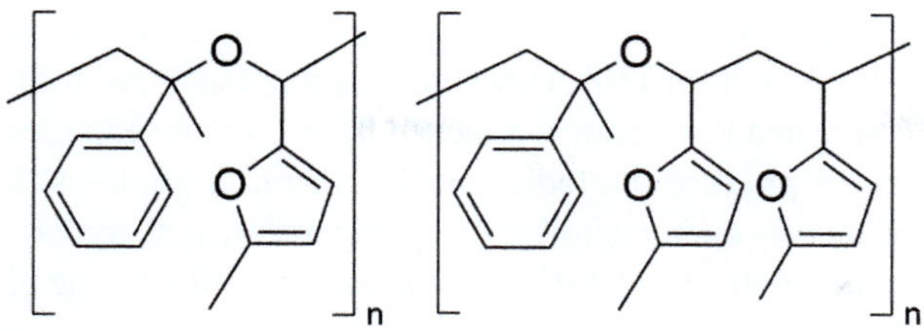

Figure 4.25 The limiting structures of MF/MST copolymers as a function of monomer feed ratios.

structure given in Figure 4.26. When MF was copolymerized with an excess of tetrahydrofuran, using methyl triflate as initiator, the alternating copolymer shown in Figure 4.27 had an Mn of about 5,000 Da.

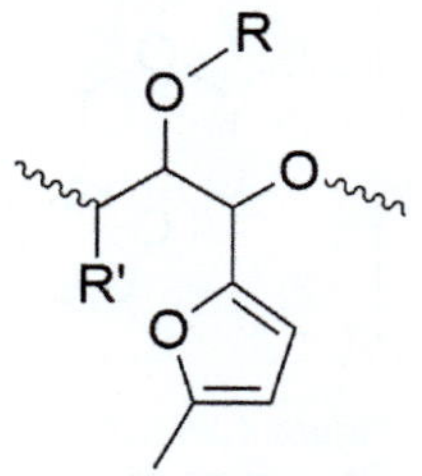

Figure 4.26 The alternating copolymer obtained in the cationic copolymerization of MF and vinyl ethers.

In all these systems, it was postulated [27, 28] that F or MF form a donor-acceptor complex with the initiator that leaves none of it free in solution. The addition of the comonomer to that solution produces one of the following alternative behavior: (i) the comonomer is incapable of competing with the furan aldehyde for the initiator because of its lower relative nucleophilicity and the result is therefore near-complete inhibition, as in case of, e.g., styrene, isobutene, and cyclopentadiene; (ii) the comonomer is less basic than the furan aldehyde, but can compete for the initiator-aldehyde complex and hence it can add to the carbonyl function of the aldehyde, thus giving alternating copolymers, as in the case of α-methyl styrene, vinyl ethers and tetrahydrofuran (the latter two only with MF); (iii) the comonomer is more nucleophilic than the furan aldehyde and "captures" the initiator, thus predominantly giving its homopolymer or copolymers with very low aldehyde-units content, as with vinyl ethers (in the case of F) and N-vinyl carbazole.

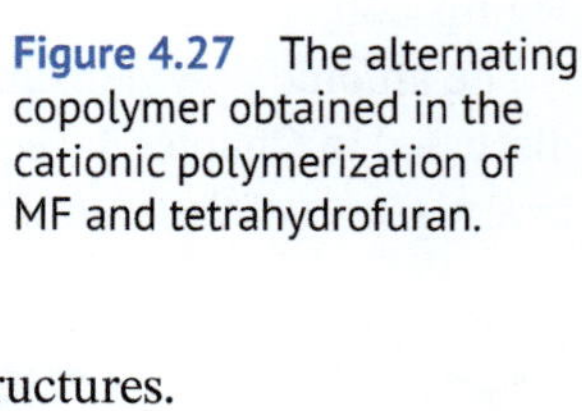

Figure 4.27 The alternating copolymer obtained in the cationic polymerization of MF and tetrahydrofuran.

The cationic polymerizations of 2-furyl vinyl ketone [10], 2-furyl oxirane [11] did not give rise to interesting materials in terms of yields and polymer structures.

4.3 Anionic Systems

The lithiation of furan is a well-known reaction that has been widely exploited in the synthesis of a variety of furan derivatives [29] without however giving an ionic species which would induce furan polymerization, whereas furan reacts with

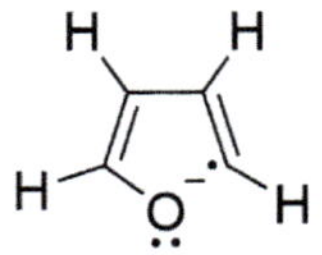

Figure 4.28 The furan radical anion generated by strong nucleophilic attacks.

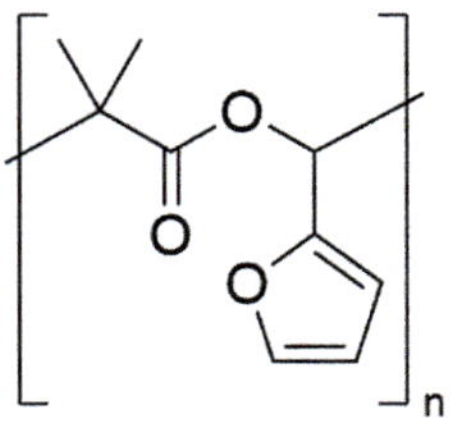

Figure 4.29 The alternating copolymer of F with dimethylketene.

sodium naphthalene to give resinous products, but no 2-furyl sodium [30]. Other metalation reactions have confirmed that furan is attacked by strong nucleophiles which produce the radical anion shown in Figure 4.28 by its ring opening, followed by uncontrolled oligomerization giving unidentified products. The same is true of 2-alkyl furans.

The behavior of 2-alkenyl furan with conventional anionic initiators is likewise disappointing since, despite previous uncorroborated claims, nBuLi, Na mirrors, or sodium naphthalene used in a variety of solvents and a wide range of temperatures, gave no polymerization with VF or MVF, suggesting that the ring was attacked in a similar way as discussed above for furan itself, rather than the vinyl moiety [31]. Surprisingly, when biphenyl sodium was tested in HMPT at room temperature, some 10% of a regular PVF (DPn ~100) was obtained, but the low yield could not be improved because of side reactions that killed that active species, even when the reaction conditions were broadly modified. To the best of our knowledge, no further report on the polymerizability of 2-alkenyl furans by anionic initiation has been published to date.

Furfural was copolymerized with dimethylketene using lithium alkyls and alcoholates to give the alternating structure shown in Figure 4.29 [32], which exhibited remarkable stereoregulatity and hence a crystalline polymer. This pioneering study was not taken up further during the succeeding 60 years!

2-Furyl vinyl ketone was polymerized using MeOK, nBuLi, and tBuOK and gave a high yield of oligomers (Mn ~1000 Da), which however turned out to contain a variety of structural units [10] suggesting that propagation could take place through the vinyl group, the furan ring and the carbonyl function, as shown in Figure 4.30.

The anionic polymerization of 2-furfurylidene methyl ketone has already been discussed in Chapter 3 (see Figure 3.6) in the context of furfural-based resins. The investigation of the anionic polymerization of 2-furan isocyanates called upon the

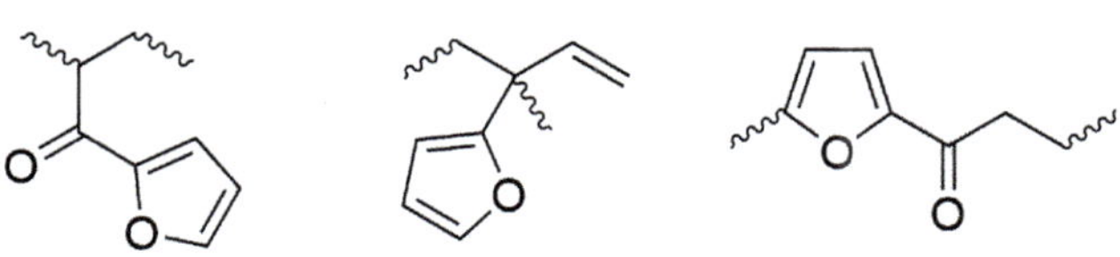

Figure 4.30 The monomer units present in the oligomers from the anionic polymerization of 2-furyl vinyl ketone.

2-furyl and the 2-furfuryl homologues (see also their synthesis and reactivity in the section on furan polyurethanes) as monomers and nBuLi or sodium naphthlalene in toluene or DMF between –60 and –80 °C [33]. The use of low temperatures and non-polar media favored linear polymerization through the opening of the N=C bond (Figure 4.31), whereas cyclic isocyanurate trimers were preferred in polar solvents (Figure 4.32). The polymers displayed a strong birefringence typical of a crystalline morphology, but the 2-furfuryl homologue (Mn >5000 Da) partly decomposed into its trimer before melting at 180–185 °C. The crystalline 2-furyl counterpart (Mn~1000 Da) exhibited a very marked thermal stability with a melting temperature of 370–375 °C without decomposition. This vast difference in thermal properties was attributed to the higher stiffness of the 2-furyl polymer, which did not readily bend to initiate the decomposition through trimer formation [33].

FO (2-Furyl oxirane) is an original furan monomer which displays an unusually high reactivity in anionic polymerizations initiated even by mild nucleophiles, whereas its free-radical and cationic polymerizations do not represent interesting systems because of a lack of reactivity and low-DP products [11]. The features of its anionic polymerization *in vacuo* were investigated in detail under a variety of conditions and using model compounds to unravel their kinetic, mechanistic and structural aspects [34–36].

The bulk polymerization initiated by KOH and NaOH at room temperature was slow and gave lower DPn values than those expected for a living anionic system [34]. A systematic study was then conducted using tBuOK in which the initial FO/initiator molar ratio was see to vary between 20 and 1,000 at temperatures between 20 and

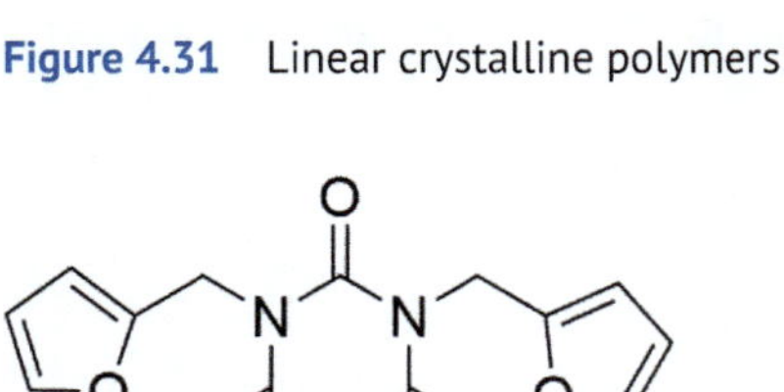

Figure 4.31 Linear crystalline polymers of the 2-furan isocyanates.

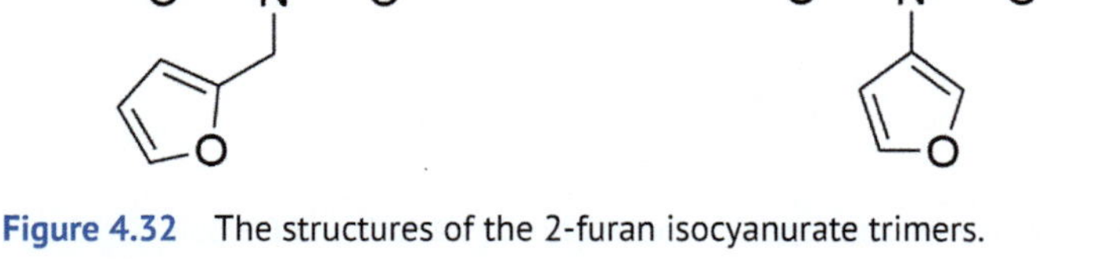

Figure 4.32 The structures of the 2-furan isocyanurate trimers.

60 °C. Polymer yields were high-to-complete within one to three days depending on the temperature, and Mn values ranged between 1,500 and 10,000 depending on the FO/initiator molar ratio. A detailed spectroscopic analysis of the polymer structure showed that the monomer insertion during propagation indicated that ring opening can occur through both the α and β modes, giving therefore amorphous polymers. Reactions in THF were slow and gave the same polymer structures, but more modest DPn. A more polar medium was therefore selected, using DMSO as solvent, which improved the polymerization rates considerably with limiting yields being reached within about 30 minutes at 20 °C [34]. The kinetic behavior of this system indicated the occurrence of a termination reaction which was the more important the lower the initiator concentration. The last initiator selected for this first study was aluminum isopropoxide, which gave fast reactions and a more stereospecific opening of the oxirane ring during the propagation steps, and hence of higher polymer tacticity [34]. The polyFO obtained here had Tg values varying between −50 (Mn = 1,000 Da) and 14 °C (Mn = 10,000 Da), and their TGA in an inert atmosphere showed the onset of thermal degradation at about 320 °C, with a 30% carbonaceous residue, typical of furan-containing polymers, as discussed in detail in another chapter. FO copolymerizations with propylene oxide (PO) failed because of the pronounced difference in monomer reactivity, which favored FO, which produced its initial homopolymerization followed by that of PO, once all the FO had been used up. Butylene oxide (BO) was found instead to display a better competitive reactivity and gave copolymers compositions from which r_1 (FO) ~ 5 and r_2 ~ 0.8 [34]. This work showed the high reactivity of FO in anionic polymerization in which two competitive initiation steps intervene through the alternative mode of ring opening, as shown in Figure 4.33. And of course, any propagation step can proceed according to the same alternative ring opening, thus giving a regioirregular polymer chain. However, with aluminum isopropoxide, the α-opening predominated by about 90% thus giving a polymer of high tacticity with the specific structure given in Figure 4.34.

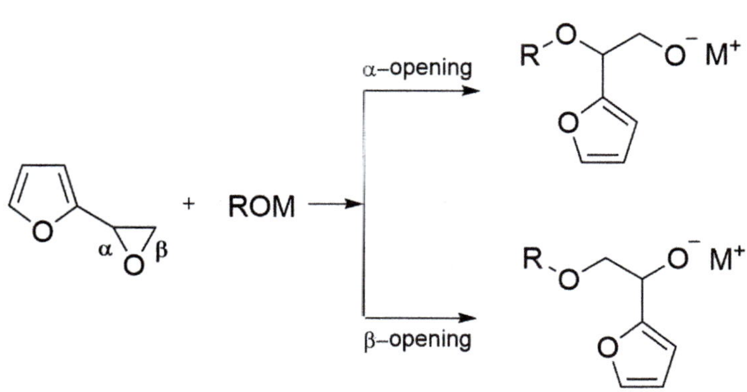

Figure 4.33 The two oxirane ring opening in the anionic initiation of FO polymerization.

As for the termination reactions, they were induced by proton abstraction either from the monomer (Figure 4.35) or from the growing polymer (Figure 4.36), giving rise to stabilized, and hence non-propagating, anions.

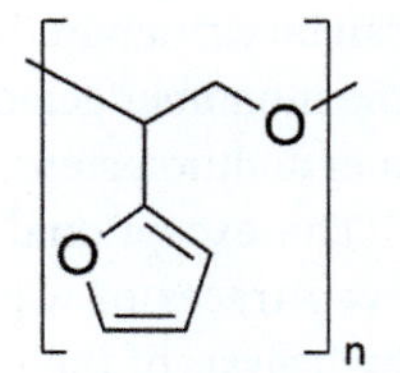

Figure 4.34 The polyFO structure arising from predominant α-ring opening.

In a second study [35], the reduction, hydrolysis, and alcoholysis of FO were examined in order to establish the origin and mechanism of its uncatalyzed OH-induced polymerization. The detailed results clearly established the initiating role of -OH groups to give FO oligomers. A remarkable extension of this conclusion was to ascertain if water could also be responsible for FO polymerization, and a series of experiments showed that thoroughly dried FO did not polymerize over on week at 80 °C, whereas the presence of millimolar concentrations of added water gave 100% of a 5,000 Da polymer within 24 h, while propylene oxide and styrene oxide were completely inert in the same conditions. It was ascertained that this process did not depend on the nature of the vessel surface and was insensitive to free-radical and nucleophilic potential inhibitors, ruling out radical and cationic mechanisms, and even the use of acidic solvents like nitromethane and carbon tetrachloride did not inhibit this polymerization, which indicated that the nature of its mechanism was not a classical anionic one. On the basis of the evidence gathered here, the steps leading to the water-induced polymerization [35]

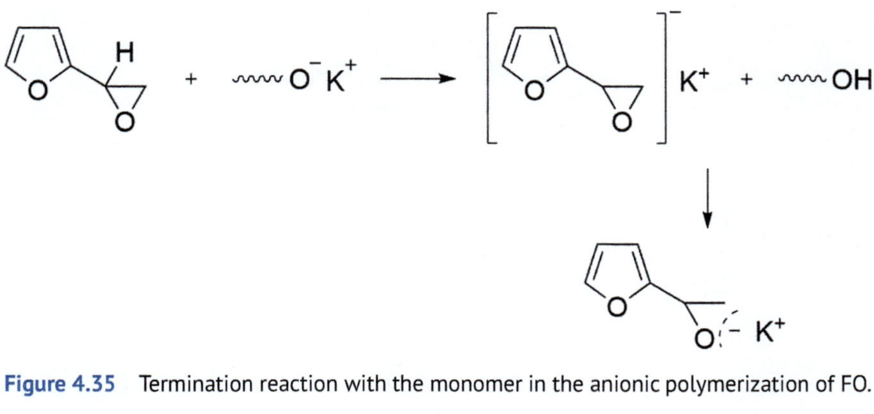

Figure 4.35 Termination reaction with the monomer in the anionic polymerization of FO.

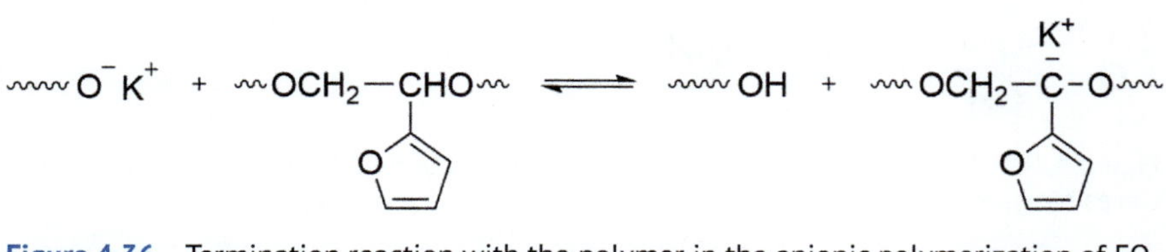

Figure 4.36 Termination reaction with the polymer in the anionic polymerization of FO.

can be written in the sequence shown in Figure 4.37, where the slow hydrolysis of the monomer generates a primary OH group responsible for the following monomer addition steps.

The exceptional reactivity of FO in the presence of mild nucleophiles and even traces of water (which in fact produces the active -OH function by the hydrolysis of the monomer) had never been encountered in any other oxiranes and opened the way to interesting macromolecular syntheses [36]. The use of simple diols and triols, as well as macrodiols as additive to FO in molar ratios of 3 to 45, produced OH-telechelic oligomers with Mn of up to 700 Da for the molecular additives, indicating that all these compounds had indeed induced the FO oligomerization, but that transfer reactions hindered the growth of higher molecular weights. An interesting application of this situation is the preparation of a fully furanic diol based on the reaction of BHMF with FO at a 1:4 molar ratio, giving the macrodiol shown in Figure 4.38 [35, 36], which was subsequently used in the synthesis of polyurethane foams, as discussed in the appropriate section.

Apart from these chain-extension reactions, grafting-from was applied to poly(vinyl alcohol) and cellulose, and again the appended chains were short because of the transfer interference. This limiting aspect requires further research in order to enhance the potential exploitation of FO as a more efficient furan-polymer modifier.

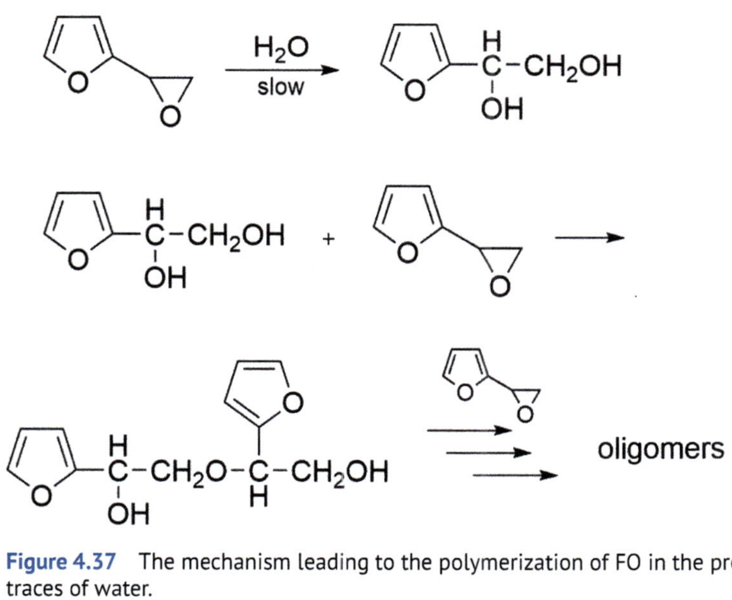

Figure 4.37 The mechanism leading to the polymerization of FO in the presence of traces of water.

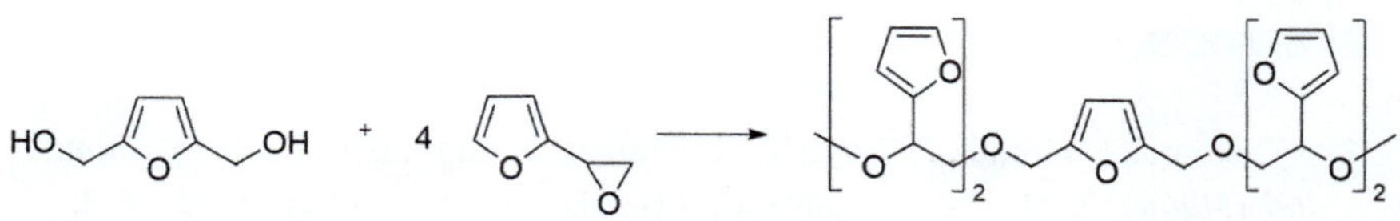

Figure 4.38 The uncatalyzed reaction of BHMF with FO at a molar ratio of 1:4.

4.4 Stereospecific Systems

Preliminary attempts to polymerize 2-alkenyl furans using typical Ziegler-Natta catalysts have produced disappointing results [37]. The use of strong catalyst combinations, like Et_3Al+$TiCl_4$, Et_3Al+$TiCl_3$ and Et_3Al+$VOCl_3$, with VF and IPF in heptane and toluene at 30–50 °C, gave low yields of insoluble and infusible polymers with FTIR spectra suggesting that the expected linear growth was marred by reactions at the C5 position and that the precipitated polymer inhibited the further initiating action of the catalysts. Weaker initiators, like $Et_2AlOtBu$+$VOCl_3$, gave similar results with VF, but reasonable yields of a soluble polymer with IPF, with an Mn higher than 6,000 Da, which however did not show any tendency to crystallize, despite spectroscopic evidence of a regular structure. Clearly, this investigation requires further research.

Alkyne furan monomers were submitted to polymerization attempts using transition metal halides, mostly $MoCl_5$, as catalysts [38], but the processes were marred by the interference of reactions between the catalyst and the heterocycle, which precluded a normal polymer growth. The synthesis of this type of polymer was therefore achieved through an indirect route, in which the furan ring was appended to a reactive polyacetylene backbone, but this rather tortuous procedure did not seem to grant further work, considering that in fact the actual furan alkyne monomers were not used in this approach.

The acquired knowledge on the furan polymers prepared by polyaddition reactions provides both deceptive and positive aspects, mostly because the unique properties of the heterocycle can lead to complex and irregular behavior, but also to original and exploitable outcomes in terms of promising materials. It is very important to place this state of the art within the context of monomers which are readily and economically prepared from furfural, a cheap renewable compound that can be readily prepared industrially from agricultural and forestry by-products available in every region of the world. This amply justifies therefore the pursuit of scientific and technological research aimed at widening the study of these polymerizations and ensuing materials. In the chapter devoted to the Diels–Alder reaction, new avenues will be highlighted concerning original furan monomers derived from furfural.

References

1 (a) Vazzana, I., Grandi, F., Hayashi, K., Munari, S., and Russo, S. (1975). *Chim. Ind. (Milan)* 57: 745–748. (b) Gaylord, N.G., Martan, M., and Deshpande, A.B. (1978). *J. Polym. Sci., Polym. Chem. Ed.* 16: 1527–1537. (c) Ragab, Y.A. and Butler, G. (1981). *J. Polym. Sci. Polym. Chem. Ed.* 19: 1175–1196.

2 (a) Fawcett, A.H., Mulemwa, J.N., and Tan, C.E. (1984). *Polym. Commun.* 25: 300–302. (b) Fawcett, A.H., Mulemwa, J.N., and Tan, C.E. (1985). *Polymer* 26: 1268–1276. (c) Chau, C.W., Fawcett, A.H., Mulemwa, J.N., et al. (1992). *Makromol. Chem.* 193: 257–273.

3 Oikawa, E., Aoki, F., and Katano, K. (1979). *Polym. J.* 11: 257–259.

4 Gandini, A. and Rieumont, J. (1976). *Tetrahedron Lett.* 25: 2101–2104.

5 (a) Gandini, A. and Belgacem, M.N. (1997). *Progr. Polym. Sci.* 22: 1203–1379. (b) Rieumont, J. (1976). Doctorate Thesis, University of Havana. (c) Rieumont, J. and Davidenko, N. (1986). *Acta Polym.* 37: 297–299. (d) Rieumont, J., Davidenko, N., and Paz, J.A. (1987). *Acta Polym.* 38: 21–23. (e) Rieumont, J., Vega, R., Davidenko, N., and Paz, J.A. (1988). *Eur. Polym. J.* 24: 909–911.

6 Sanchez, S.O., Marra, F., Dibenedetto, A., Aresta, M., and Grassi, A. (2014). *Macromolecules* 47: 7129–7137.

7 Dillingham, K.A. (1994). Ph. D. Thesis, University of Lancaster, UK.

8 (a) Hernandez, C., Galego, N., Llerandi, N., and Gandini, A. (1974). *Revista CENIC* (Cuba) 5: 217–226. (b) Romani, F., Corrieri, R., Braga, V., and Ciardelli, F. (2002). *Polymer* 42: 1115–1131. (c) Coiai, S., Passaglia, E., Aglietto, M., and Ciardelli, F. (2004). *Macromolecules* 37: 8414–8423.

9 (a) Bevington, J.C. and Harris, D.O. (1967). *Polym. Lett.* 5: 799–802. (b) Peniche, C., Zaldivar, D., Bulay, A., and San Roman J. (1993). *Polym. Degrad. Stab.* 40: 287–295. (c) Peniche, C., Zaldivar, D., Bulay, A., and San Roman J. (1993). *J. Polym. Sci. Polym. Chem. Ed.* 31: 625–631. (d) Peniche, C., Zaldivar, D., Bulay, A., and San Roman J. (1992). *Polymer* 33: 4625–4629. (e) Peniche, C., Zaldivar, D., Bulay, A., and San Roman J. (1993). *J. Appl. Polym. Sci.* 50: 2121–2127. (f) Zaldivar, D., Peniche, C., Bulay, A., and San Roman, J. (1993). *Biomaterials* 14: 1073–1079. (g) Peniche, C., Zaldivar, D., Gallardo, A., and San Roman J. (1994). *J. Appl. Polym. Sci.* 54: 959–968. (h) Zaldivar, D., Davidenko, N., Peniche, C., et al. (1994). *Polymer*, 35: 2390–2396.

10 Labidi, A., Salon, M.-C., Gandini, A., and Cheradame, H. (1985). *Polym. Bull.* 14: 271–278.

11 Salon, M.-C., Amri, H. and Gandini, A. (1990). *Polym. Comm.* 31: 210–212.

12 Yoshida, N., Kasuya, N., Haga, N., and Fukuda, K. (2008). *Polym. J.* 40: 1164–1169.

13 Han, M., Liu, X., Zhang, X., et al. (2017). *Green Chem.* 19: 722–728.

14 Apellaniz, J.E. (1975). US Patent 3881935A, to Konica Minolta Graphic Imaging USA Inc.

15 Ishigaki, A. and Shono, T. (1974). *Bull. Chem. Soc. Jpn.* 47: 1467–1470.
16 (a) Armour, M., Davies, A.G., Upadhyay, J., and Wassrmann, A. (1967). *J. Polym. Sci.* A15: 1527–1538. (b) Kresta, J. and Livingston, H.K. (1970). *Polym. Lett.* 8: 795–803.
17 Shindo, K. and Ishikawa, S. (1977). *J. Chem. Soc. Jpn., Ind. Chem.* 2: 245–249.
18 Lamb, B.S. and Kovacic, P. (1980). *J. Polym. Sci., Polym. Chem. Edn.* 18: 2423–2436.
19 (a) Alvarez, R., Gandini, A., Martinez, R., Ortiz, P.J., and Pérez, C.S. (1974). *Rev. CENIC* (Cuba) 5: 179–193. (b) Gandini, A. and Martinez, R. (1983). *Rev. Cienc. Quim.* (Cuba) 14: 111–115.
20 Gandini, A. and Martinez, R. (1976). *J. Polym. Sci. Symp.* 56: 79–90.
21 (a) Gandini, A. and Martinez, R. (1983). *Makromol. Chem.* 184: 1189–1211. (b) Martinez, R., Alvarez, R., and Delaguardia, C. (1995). *Polym. Bull.* 34: 57–62.
22 Alvarez, R., Gandini, A., and Martinez, R. (1982). *Makromol. Chem.* 183: 2399–2413.
23 (a) Gandini, A. and Martinez, R. (1985). *Rev. Cienc. Quim.* (Cuba) 16: 275–288. (b) Martinez, R and Gandini, A. (1988). *Rev. Cubana Quim.* 4: 39–45. (c) Martinez, R., Alvarez, R. and Gandini, A. (1989). *Acta Polym.* 40: 613–618.
24 (a) Salon, M.-C. (1984). *Doctorate Thesis*, Grenoble National Polytechnic Institute, France. (b) Razzouk, H., Bouridah, K. Cheradame, H., and Gandini, A. (1984). In: *Cationic Polymerization and Related Processes*. (ed. E.J. Goethals). Academic Press, New York, 355–362.
25 (a) Salon, M.-C. (1984). *Doctorate Thesis*, Grenoble National Polytechnic Institute, France. (b) Gandini, A. and Salon, M.-C. (1990). *US Patent 5260378-A* to the DSM Co. (Holland). (c) Gandini, A., Salon, M.-C., Choura, M., et al. (1992). *Makromol. Chem, Macromol. Symp.* 60: 165–176. (d) Hadjikyriacou, S. and Faust, R. (1999). *Macromolecules* 32: 6393–6399. (e) Faust, R. (2000). *Macromol. Symp.* 157: 101–108. (f) Kwon, Y. and Faust, R. (2004). *Adv. Polym. Sci.* 167: 107–135. (g) Lange, A., Rath, H.P., and Lang, G. (2004). *Macromol. Symp.* 215: 209–214.
26 Spange, S., Hohne, S., Franche, V., and Gunther, H. (1999). *Macromol. Chem. Phys.* 200: 1054–1061.
27 Gandini, A. and Rieumont, J. (1977). *Brit. Polym. J.* (3): 28–35.
28 Brunache, M.-H. (1984). Doctorate Thesis, National Polytechnic Institute, Grenoble, France.
29 (a) Ramanathan, V. and Levine, R. (1962). *J. Org. Chem.* 27: 1216–1219. (b) Comins, D.L. and Killpack, M.O. (1987). *J. Org. Chem.* 52: 104–109. (c) Feringa, B.L., Hulst, R., Rikers, R., and Brandsma, L. (1988). *Synthesis* (4): 316–318.
30 Normant, H. and Angelo, B. (1961). *Bull. Soc. Chim. Fr.* 10: 1988–1992.
31 Gandini, A. and Hernandez, C. (1978). *Polym. Bull.* 1: 221–225.
32 Natta, G., Mazzanti, G., Pregaglia, G.F., and Pozzi, G. (1962). *J. Polym Sci.* 58: 1201–1210.
33 Hui, Z. and Gandini, A. (1991). *Polym. Bull.* 26: 383–390.

34 Amri, H., Belgacem, M.N., and Gandini, A. (1996). *Polymer* 37: 4815–4821.
35 Amri, H., Belgacem, M.N., Signoret, C., and Gandini, A. (1996). *Polym. Int.* 41: 427–435.
36 Boufi, S., Belgacem, M.N., and Gandini, A. (1997). *Polym. J.* 29: 479–486.
37 Gandini, A. and Hernandez, C.D. (1981). *Makromol. Chem. Rapid Commun.* 2: 351–354.
38 Méleares, C., Costa, G., Gandini, A., and Valenti, B. (1997). *Acta Polym.* 48: 21–24.

5

Polymers from Polycondensation (Step) Reactions

5.1 Polyesters

The class of macromolecules known as polyesters possess repeat carboxylate ester groups in their backbones, excluding all polymers bearing ester *pendant* groups, like acrylic and methacrylic polymers, poly(vinyl esters), and esterified polysaccharides. The synthetic methods that are generally used for the preparation of linear polyesters may involve (i) the stoichiometric reaction between dicarboxylic acids and glycols, which may lead to undesired glycol condensation and/or dicarboxylic acid decomposition at high temperatures; (ii) the stoichiometric transesterification reaction between diesters and glycols, which is preferable considering the lower amount of energy needed to remove the alcohol – instead of water – formed during condensation and in the case of high melting and poorly soluble dicarboxylic acids; (iii) the use of hydroxyacids, like *p*-hydroxyethoxybenzoic acid, which are capable of self-condensation to form polyesters; (iv) the reaction of dicarboxylic acid anhydrides with glycols; (v) the reaction of acid chlorides with diphenols, which is quite efficient when a scavenger of HCl, such as tertiary amines, is added to the aqueous phase; (vi) the ring opening of lactones, among other synthetic routes that can be found in the literature but are not in common use [1].

Poly(ethylene terephthalate) (PET, Figure 5.1) is produced from terephthalic acid and ethylene glycol via polycondensation, and is considered the most important commercial polyester, with an annual market size estimated in 50–60 Mt. At the present time, approximately 60% of PET are applied in textile fibers, 30% in beverage bottles, and 10% in jars, containers and packaging [2].

The increasing interest in the development and consolidation of integrated biorefineries has allowed the viable production of important bio-based building blocks that can be used for the synthesis of conventional and original polyesters, in a strategy that not only offers sustainable solutions for conventional petroleum-based polyesters, such as PET, poly(propylene terephthalate) (PPT) and poly(butylene succinate) (PBS), but also new insights for the development of original ones [3].

Furan Polymers and their Reactions, First Edition. Alessandro Gandini and Talita Martins Lacerda.

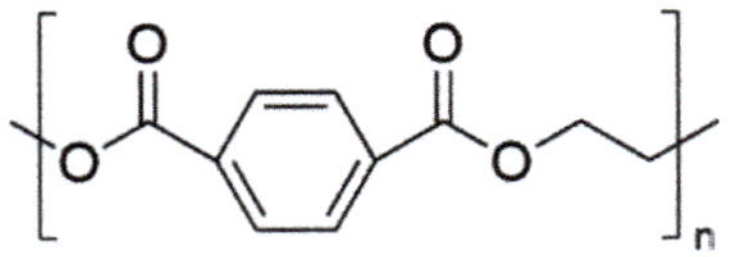

Figure 5.1 Chemical structure of poly(ethylene terephthalate) (PET).

Recently, a drive toward bio-based production of PET has been motivated, on the one hand, by its wide use in the food and beverage packaging, and on the other hand, by the need of reducing the dependence of the plastic industry on the fossil platform. Some promising catalytic processes to convert lignocellulosic biomass into ethylene glycol and terephthalic acid for producing PET are available in the literature [4], although a non-neglectable number of challenges, mostly related to low reaction yields, still need to be addressed for most of them to reach industrial significance. Coca-Cola has been a leading Figure in this context in recent years, with a first initiative related to the production of PET bottles using sugarcane-based ethylene glycol and fossil-based terephthalic acid; more recently, a 100% bio-based prototype was made from corn-based terephthalic acid [5].

In this context, the furan platform opens the way for an *even wider* range of possibilities for plant-based polyesters: many differently bridged mono- and difuran dicarboxylic acids synthesized from furfural (F) and 5-hydroxymethylfurfural (HMF) may be reacted with a comprehensive series of aliphatic and/or aromatic diols to produce fully or partially bio-based thermoplastic polyesters with tunable properties; unsaturated and/or polyfunctional alcohols may be also considered for the synthesis of polyester resins, which are useful materials with applications in the construction and automotive industries, as well as for wood and furniture finishing.

Such huge range of possibilities has given rise to a very abundant literature on furan polyesters [6–16], with dozens of novel contributions being published yearly since the beginning of the last decade (Figure 5.2). A pioneering investigation [17] initiated this topic by describing the synthesis and thermal properties of polyesters from FDCA and 2,5-bis(hydroxymethyl)furan or 1,6-hexanediol via interfacial or solution methods, but was not cited by other researchers directly working on the topic before the second half of the 1980s, which is not surprising considering that the characterization of these polyesters was relatively limited because of the lack of a complete spectroscopic assessment, and of the modest information that could be obtained on their degrees of polymerization (DPs) [7].

The first significant wave of publications on furan polyesters was witnessed during the 1990s, mainly due to the efforts of a prolific research group [18], who studied polyesters from difuran dicarboxylic acids synthesized from F; the second relevant contribution came after 2009, when another group working under the same leader successfully produced the furan counterpart of PET, poly(ethylene 2,5-furandicarboxylate) (PEF) [19].

Herein, furan polyesters will be presented in two separate families of macromolecular structures, namely those based on monomers incorporating a difuran core

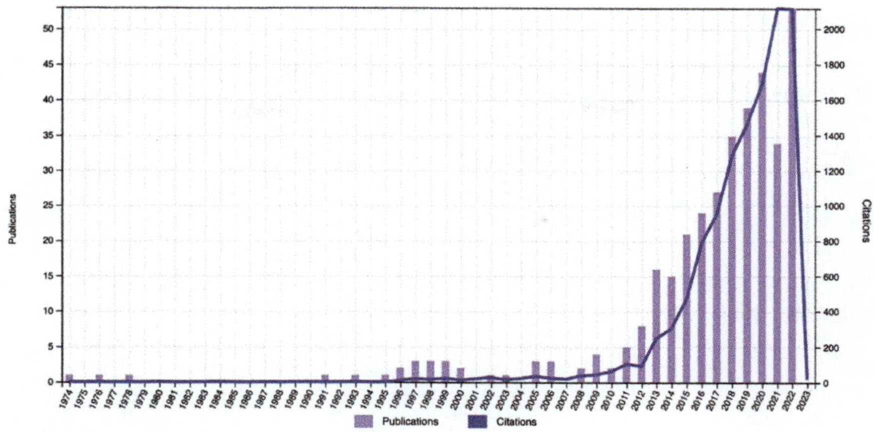

Figure 5.2 Evolution of the number of publications and ensuing citations using the keywords "furan" and "polyester". *Source:* Adapted from ISI - Web of Knowledge.

(mostly – but not exclusively – based on F), and those based on monomers incorporating a single furan core, which is the case of 2,5-furan dicarboxylic acid (FDA) and its corresponding hydroxyacids. We decided to start from the former, as a strategy that allows the design of original furan-based polyesters; the latter, however, is much more intense in terms of qualitative and quantitative publication indicators, mostly because of the development related to research on HMF and, in particular, on FDCA [20]. It will be, therefore, tackled in a further section.

5.1.1 Polyesters from Monomers Bearing a Difuran Core

Furan undergoes condensation with aldehydes and ketones to give secondary and tertiary alcohols (Figure 5.3), which may further react with furan to form oligomers [21]. This reaction has been used as a versatile tool for the synthesis of bifunctional monomers to be incorporated in polycondensates.

The original series of papers mentioned above [18] aimed at describing furan polyesters based on key diesters prepared by the condensation of aldehydes or ketones with 2-ethyl furoate in the presence of sulfuric acid as catalyst (Figure 5.4). The difuran diesters (or the corresponding diacid chlorides) were reacted with a variety of diols to produce polyesters (Figure 5.5), whose synthesis, physical characterization, and properties were thoroughly described.

During the last 10–15 years, the strategy kept being pursued by different research groups who (i) exploited varying combinations of the aforementioned monomers to prepare homo- and copolyesters, (ii) designed original monomer architectures, mainly considering variations on the bridges between the two furan rings, and (iii) designed and optimized original reaction pathways that could lead to a viable large-scale production of such monomers. Table 5.1 summarizes some

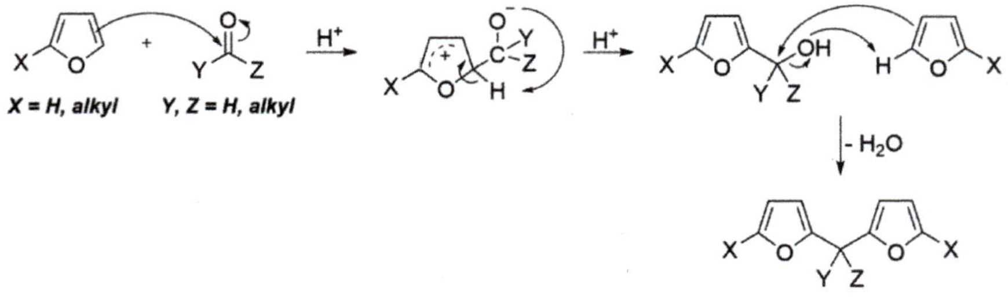

Figure 5.3 Condensation reaction of furan and 2-monosubstituted furans with aldehydes and ketones.

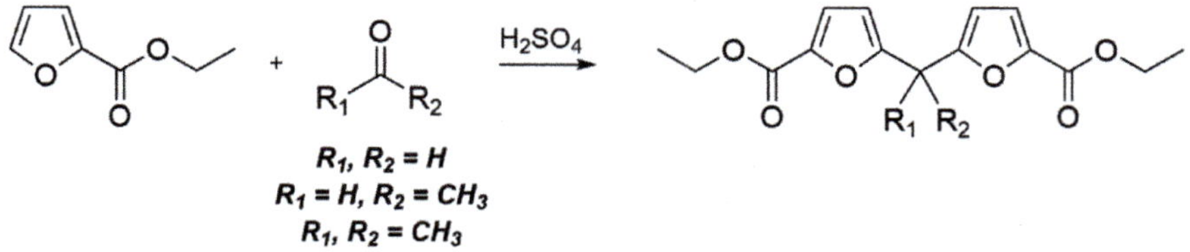

Figure 5.4 Synthesis of difuran monomers.

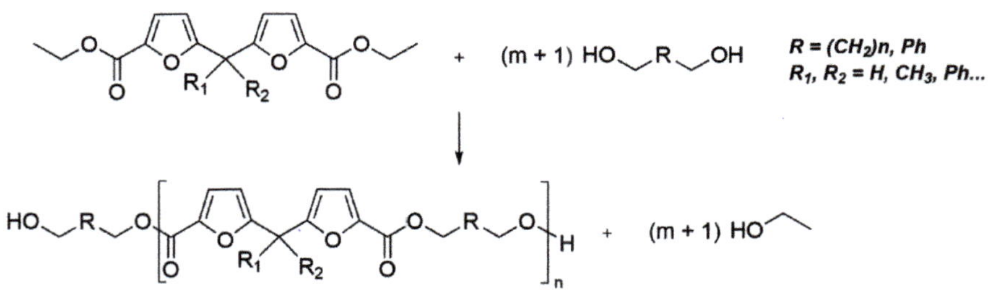

Figure 5.5 A set of furan polyesters prepared by transesterification using difuran diesters and diols.

of the most relevant initiatives published so far and is followed by the specific technical information.

In the light of the first investigations on the synthesis of polyesters based on the difuran diesters sketched in Figure 5.4, other interesting contributions involved the use of comonomers such as isosorbide [22], glycols of varying carbon chain lengths, diamines and a sulfonated benzene ring [23–25], which allowed the design of furan-based polyesters and poly(ester amide)s with acceptable thermal and mechanical properties. The same difuran core, prepared from the condensation of ethyl furan-2-carboxylate with acetone in an acidic medium, was also submitted to a further reaction step with propargyl alcohol, and a bifunctional terminal alkyne was obtained and used to synthesize a thermally stable poly(ester

Table 5.1 Summary of some representative difuran monomers and their corresponding polyesters.

Molecular structure of the difuran monomer	Other (co)monomer(s) used for polymer synthesis	Final product(s)	Reference
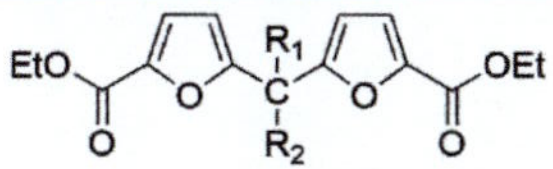 $R_1 = CH_3$ or nC_3H_7 $R_2 = CH_3$ or C_6H5 or nC_3H_7	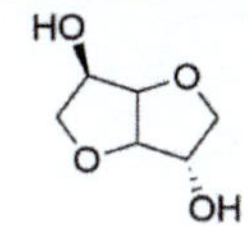	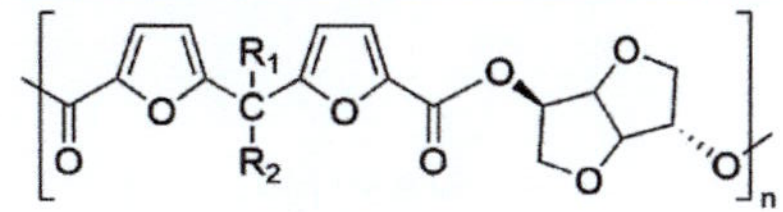 polyesters	[22]
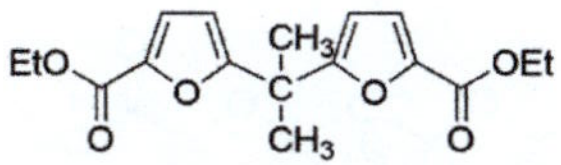	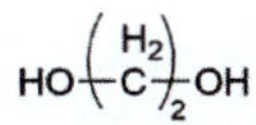 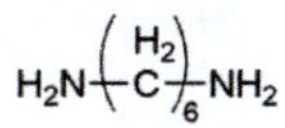	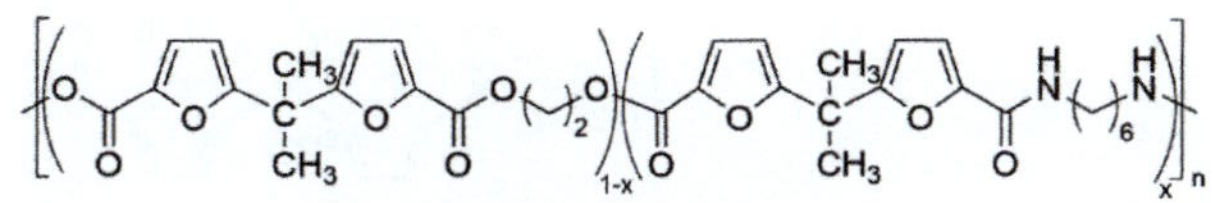 poly(ester amide)s	[23]
	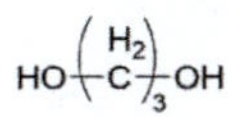 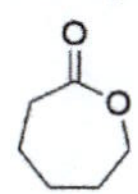	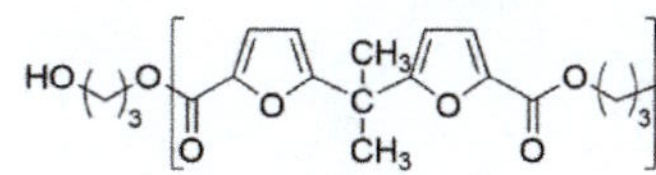 pre-polymer 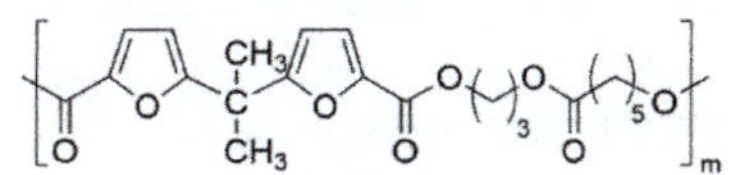 polyesters	[24]
	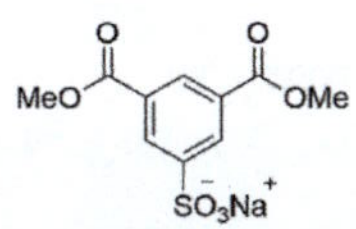 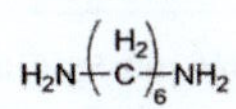 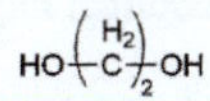	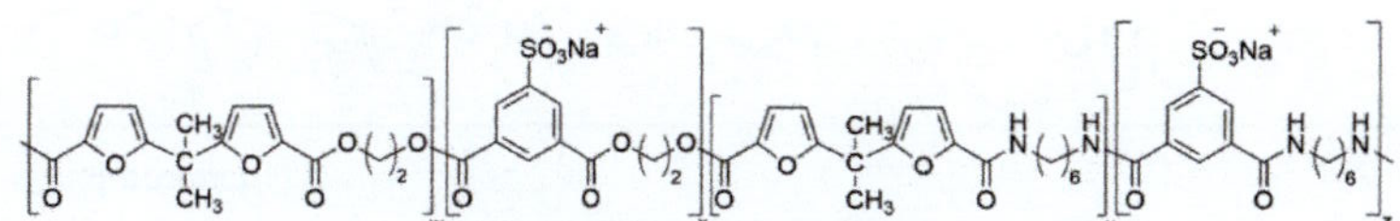 sulfonated poly(ester amide)s	[25]

(*Continued*)

Table 5.1 (Continued)

Molecular structure of the difuran monomer	Other (co)monomer(s) used for polymer synthesis	Final product(s)	Reference
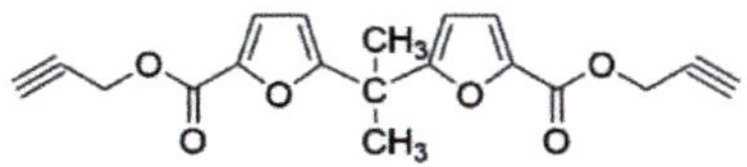	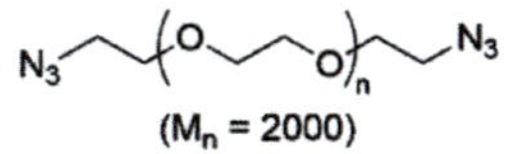	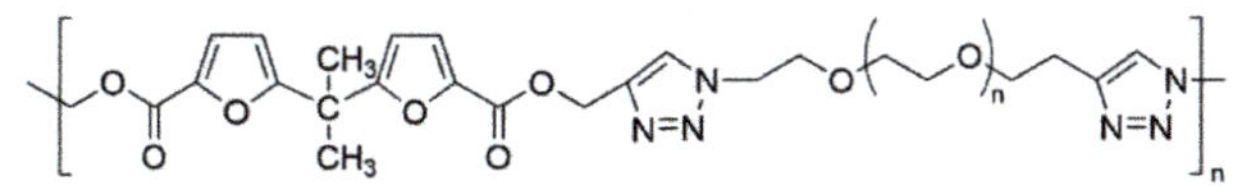 poly(ester triazole)	[26]
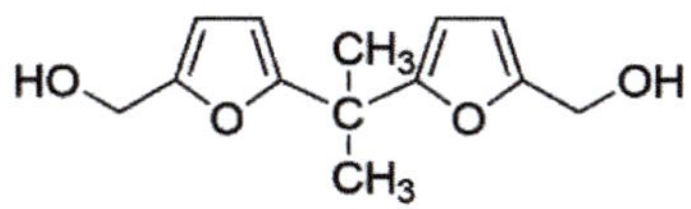	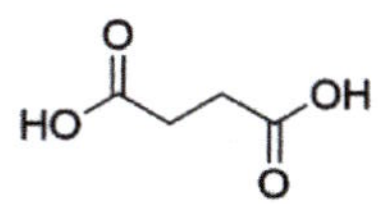	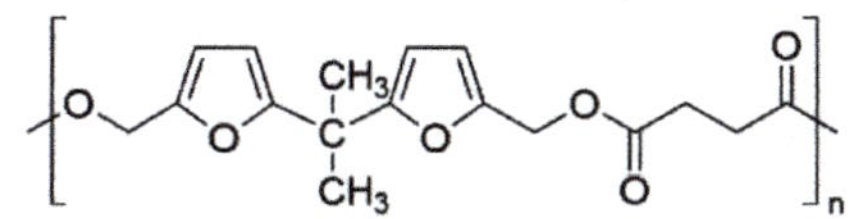 polyester	[27]
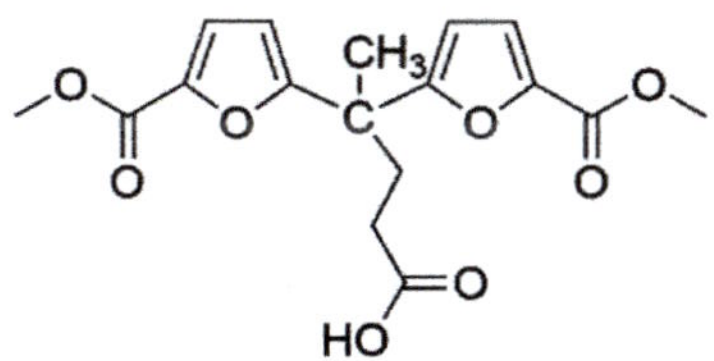	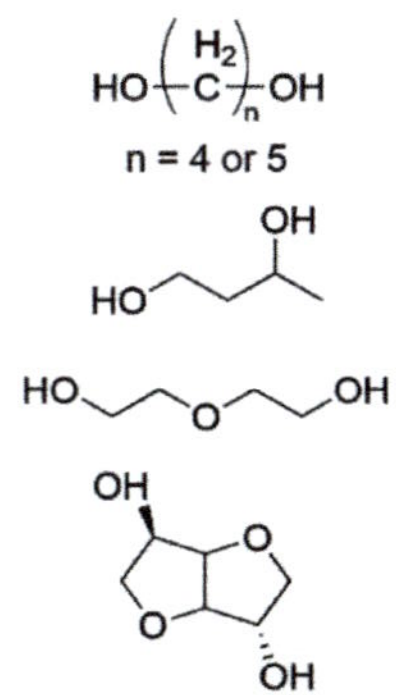	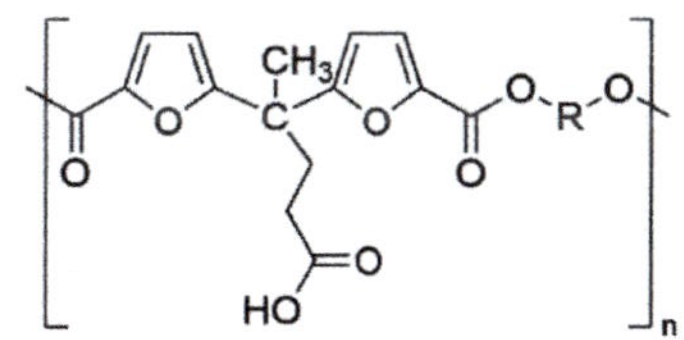 polyesters	[28]

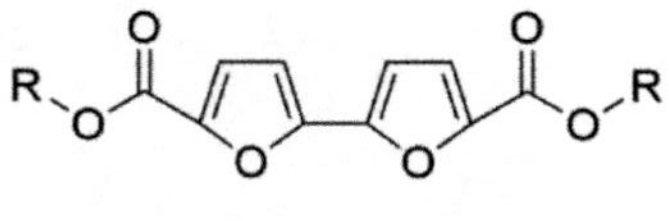 R = CH_3 or CH_2CH_3	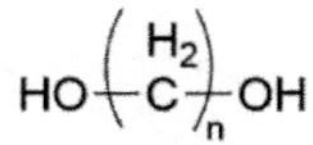 n = 2 to 6	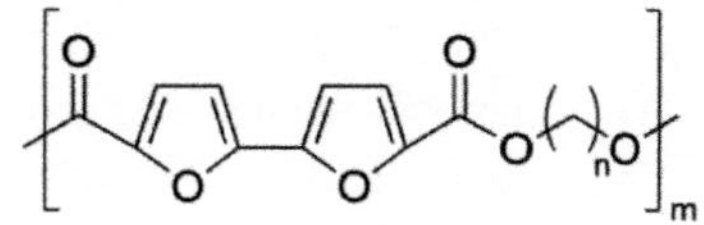 polyesters	[29,30]
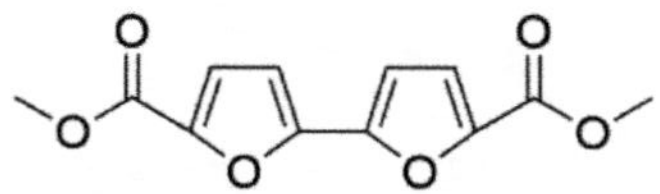	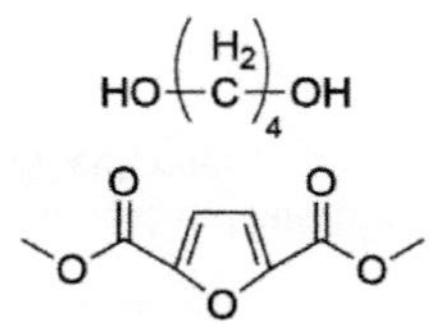	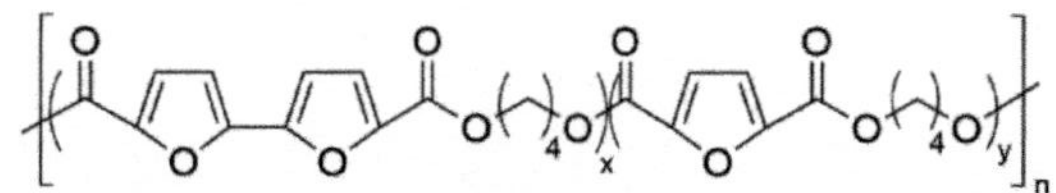 polyesters	[31]
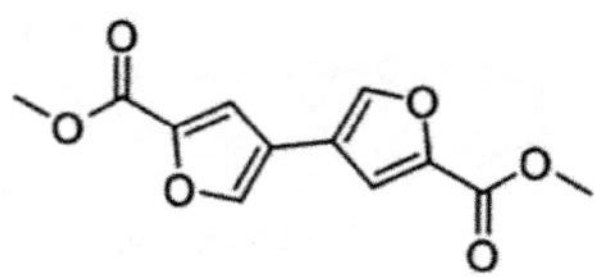		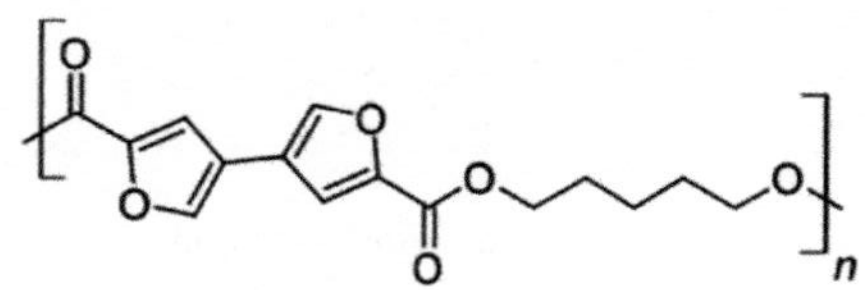 polyesters	[32]
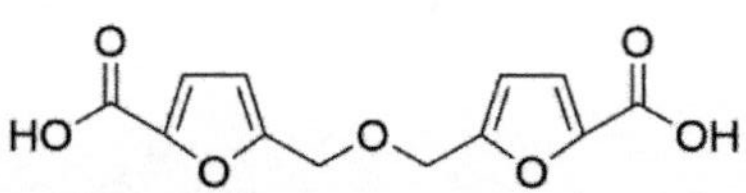	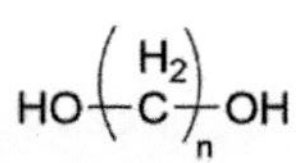 n = 2 or 4	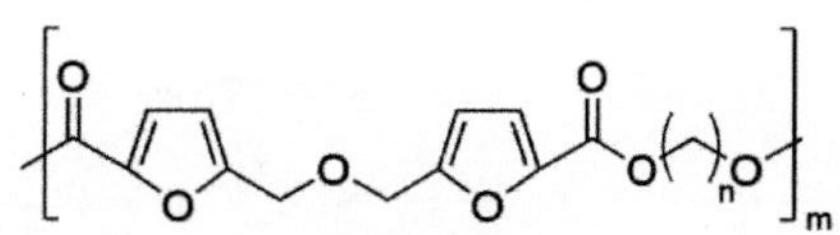 polyesters	[33]

(Continued)

Table 5.1 (Continued)

Molecular structure of the difuran monomer	Other (co)monomer(s) used for polymer synthesis	Final product(s)	Reference
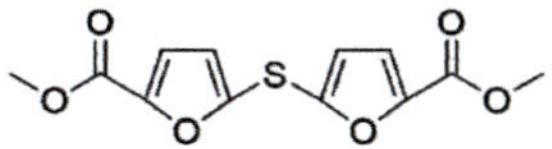	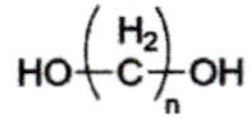 n = 2 to 5	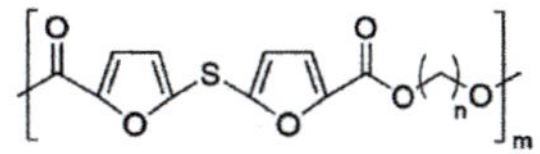 poly(thioester)s	[34]
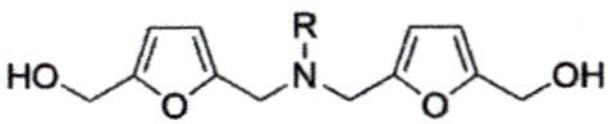 R = CH_2-Ph; CH_2-CH_2-Ph; CH_3-CH_2; $(CH_2)_6CH_3$; $(CH_2)_{11}CH_3$	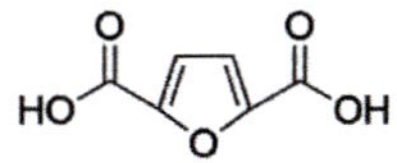	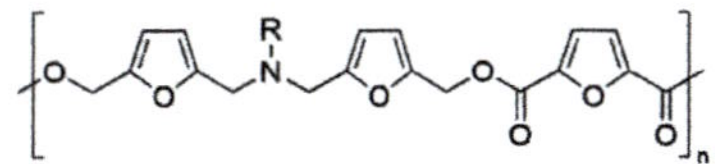 poly(ester amine)s	[35]
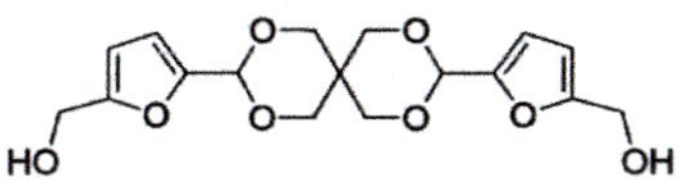	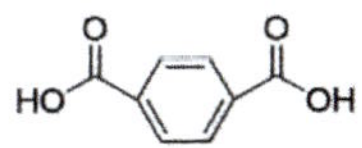 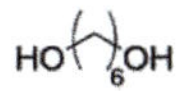	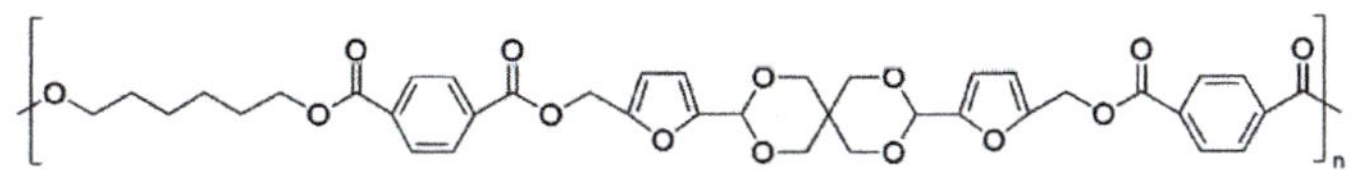 polyesters	[36]
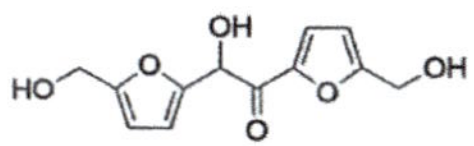 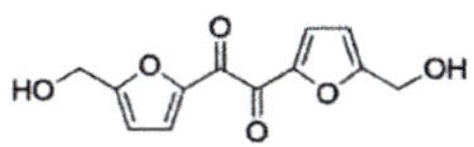 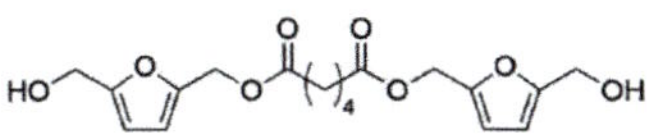	saturated and unsaturated acyl chlorides	linear and crosslinked polyesters and poly(ester urethanes)	[37]
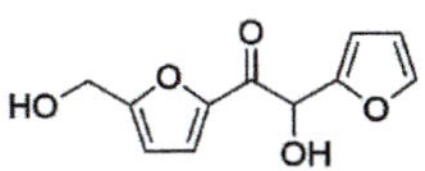 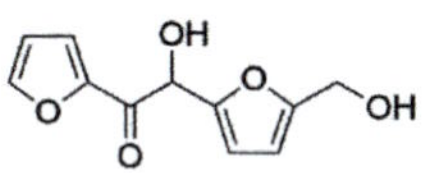	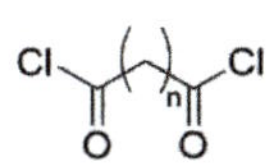 n = 2, 4 or 6	linear polyesters and polyurethanes bearing pendant furan rings	[38]

triazole) via azide-alkyne "click" reaction catalyzed by Cu(I) [26]. Polymers were isolated with molecular weights ranging from 7 to 11 kDa and with glass transition temperatures varying from −35 to −19 °C.

Efforts on the establishment of efficient routes to produce such difuran molecules as building blocks for various furan-based renewable polymeric materials mean a step forward towards industrially relevant materials. In this context, a common difuran dibromide intermediate synthesized from commercially available 2-methylfuran was proven to be readily converted to analogous difuran diacid, diol, and diamine compounds in high to quantitative isolated yields [27]. A supposedly easily up-scalable synthetic route was initiated by the protection of the aldehyde functionality of commercial furfural, followed by Friedel-Crafts alkylation, to yield a furan dimer; two further steps of aldehyde deprotection and reduction led to a difuran diol, that was polymerized with succinic acid [29].

More recently, monomers bearing difuran cores slightly different in terms of molecular structure started to receive attention for the preparation of polyesters. The condensation reaction of 2-methylfuroate with levulinic acid produced a fully bio-based diester [28], which was polymerized with 1,3-, 1,4-, and 1,5-butanediol, diethyleneglycol, and isosorbide to give polyesters with good thermal stabilities and glass transition temperatures (Tg) ranging from 37 to 119 °C. Although the authors did not consider this possibility, the pendant carboxylic functionality of levulinic acid could be exploited to prepare polyester network with tunable properties.

A 2,2′-bifuran-5,5′-dicarboxylate was synthesized for the first time by reductive or oxidative homocoupling of furan carboxylate derivatives in the presence of a nickel or palladium catalyst [30]. Another group reported the preparation of the same monomer, although using a different route [31,32]. Both studies demonstrated the viable utilization of this original building block for the synthesis of polyesters exhibiting good barrier properties, i.e., low permeability to oxygen, moisture, and ultraviolet (UV) light. An outstanding oxygen barrier profile was reached for polyester prepared from a 3,3′-bifuran-5,5′-dicarboxylic acid [39]. It is worth mentioning that, in this particular case, the authors investigated the aging of samples under air for 3 to 10 weeks, which led to insoluble materials in common organic solvents (including hexafluoroisopropanol) with improved gas permeability properties. The behavior is explained by crosslinks formed in air, although a tentative mechanism was not proposed.

Polycondensation monomers bearing two furan rings separated by oxygen, sulfur and nitrogen bridges were prepared and used for the synthesis of polyesters [33], poly(thioester)s [34], and poly(ester amine)s [35]. In the case of the sulfide-bridged ones, the authors achieved oxygen barrier properties 1.9 to 11.2 times higher than those observed for amorphous poly(ethylene terephthalate), as well as excellent thermal stability, good mechanical properties and a slight UV shielding effect. Outstanding properties were achieved when difuranyl monomers-bis-(hydroxymethylfurfuryl) amines [35b] bearing varying side chains were reacted with 2,5-furan dicarboxylate [35], which gave rise to poly(ester amine)s whose Tgs varied from ~3 to 50 °C, tensile moduli from ~3 to 550 MPa, and elongations at break reaching 1100%. One of the polyesters shielded approximately 100% UV light, and another one reduced gas permeability coefficients (PCO_2 and PO_2) by approximately three times when compared with those of PET. A peculiar difuran diol was prepared via a one-step acid-catalyzed acetalation of HMF and pentaerythritol [36], which was associated with 1,6 hexanediol and dimethyl terephthalate for the synthesis of polyesters. The so-called spirocyclic core enabled increased Tgs, but the high sensitivity to heat of the diol limited its incorporation into the final materials.

Homo- and cross-coupling reactions of F and HMF have been described, and difuran monomers bearing one, two and three hydroxyl groups were used to synthesize linear and crosslinked polyesters and poly(ester urethanes), some of which bore pendant furan rings that were used to form thermally reversible crosslinked networks via the Diels–Alder reaction (see Section 6) with a jeffamine-like bismaleimide [37,38].

5.1.2 Polyesters from Monomers Bearing a Single Furan Core

The investigations described above comprise *most* studies related to the use of difunctional furan monomers for the synthesis of polyesters, although some may have been inevitably left out. Because of their permanent validity, however, in the case of single-furan-based monomers, the same strategy is virtually impossible to follow: the extremely lively interest in the preparation of HMF as a promising platform chemical derived from renewable resources, and the surge of its direct derivative, 2,5-furan dicarboxylic acid (FDCA) (Figure 5.6), have been leading to impressive amounts of publications annually including books, book chapters, patents, scientific articles, etc. In 2015, an extensive review covered the most relevant contributions made to the development of FDCA-based polymers up to that moment [40], and although many other initiatives have been reported to significantly increase the volume of data generated so far, the general idea used by the authors to present this large class of macromolecules remains valid, and is related to the molecular structure of the diol associated with FDCA for the synthesis of polyesters. Herein, the focus will be limited to more recent investigations, respecting a similar logic: because of its unicity in terms of academic and industrial relevance, poly(ethylene 2,5-furandicarboxylate) (PEF) will be presented first, followed by the other representatives of the family of linear aliphatic analogues, i.e., those that bear larger methylene sequencies. The strategies involving the use

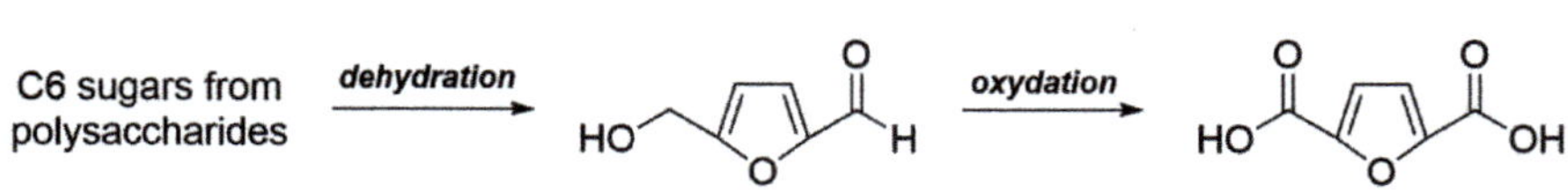

Figure 5.6 General chemical route to synthesize FDCA from bio-based sugars through HMF as a value-added intermediate.

of cyclic aliphatic or aromatic diols will also be tackled, as well as their corresponding copolymers and blends, considering, however, that a deliberate and inevitably subjective selection will be made in the choice of papers referenced within the vast amount of material available.

5.1.2.1 Poly(ethylene 2,5-furandicarboxylate)

The first systematic approach within this topic was carried out using FDCA or its diacid derivatives (as well as 2,3-, 2,4- and 3,4-homologues), and bis(hydroxymethyl) furan (BHMF) (Figure 5.7) reacting with aliphatic or aromatic counterparts, or between themselves to give the fully furanic structures (Figure 5.8) [7,17]. The ensuing polyesters had relatively low DP values, as suggested by their low intrinsic viscosities. Among the numerous structures studied, only those bearing 2,5-enchainments exhibited a clear melting feature on DSC thermograms [7].

After almost three decades of latency, the application of a patent in 2006 by Canon Kabushiki Kaisha on the preparation of three FDCA-based polyesters, i.e.,

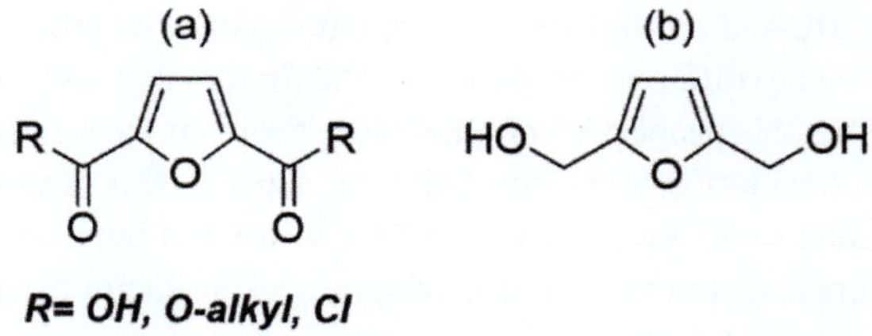

Figure 5.7 Furan building blocks considered for the synthesis of polyesters. (a) 2,5-furan diacid derivatives; (b) bis(hydroxymethyl)furan (BHMF).

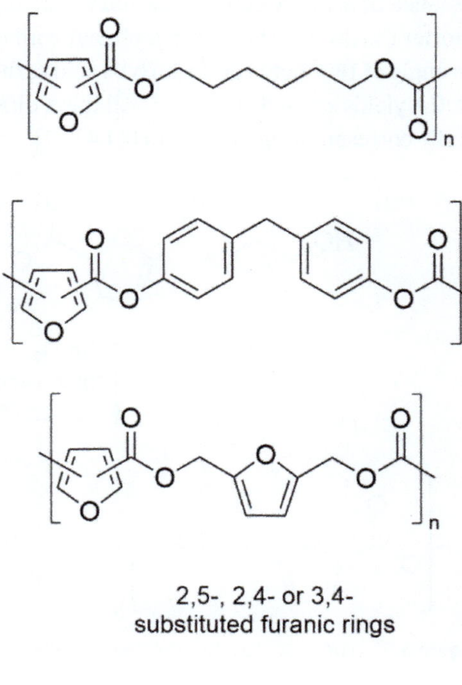

Figure 5.8 General representation of 2,5-, 2,4- and 3,4- substituted furan polyesters.

poly(ethylene-2,5-furandicarboxylate) (PEF), poly(propylene-2,5-furandicarboxylate) (PPF) and poly(butylene-2,5-furandicarboxylate) (PBF) [41], and of a second one by Mitsubishi on similar materials [42], were followed by a thorough investigation on the synthesis of poly(ethylene-2,5-furandicarboxylate) (PEF) in 2009 (Figure 5.9) [19]. On that occasion, the authors proposed three different routes for the synthesis of PEF, i.e., (i) solution polymerization of chlorinated FDCA and ethylene glycol, (ii) interfacial polymerization of chlorinated FDCA and ethylene glycol, and (iii) esterification of FDCA and ethylene glycol in the presence of hydrochloric acid, followed by the transesterification of the produced diesters under reduced pressure and using antimony (III) oxide as catalyst (Figure 5.9) [19,43]. The latter investigation described PEF as a semicrystalline polyester, displaying a glass transition (Tg) at 75–80 °C, a melting temperature (T_m) of 210–215 °C, and thermal stability up to ca. 300 °C [19,44]. In terms of mechanical properties, studies carried out at the first half of the 2010s indicated Young's moduli of 2070–2450 MPa, a maximum stress range of 35–67 MPa and an elongation at break of 3–4% [40,44,45].

Variations on the choice of the specific furan electrophile (FDCA, chlorinated FDCA, dimethyl 2,5-furandicarboxylate) and other reaction parameters obviously lead to different properties of the final materials [19,46]. For that reason, comparing the properties of PEF with those of PET is not always a trivial task [43]. Considering some fundamental aspects, the values of delocalization energy (22 kcal mol^{-1} vs. 36 kcal mol^{-1} for furan and benzene, respectively) clearly indicate once again that the furan ring has a significant diene character, resulting in a different chemical reactivity. Moreover, terephthalic acid exhibits an interatomic distance of 5.731 Å between carboxylic acid groups, while in FDCA it is notably shorter (4.830 Å). The linear *p*-phenyl connection in terephthalic acid results in an angle of 180° between carboxylic acid carbons, while the nonlinear structure of FDCA yields an angle of 129.4°. All these differences result in different properties of the corresponding polyesters [44,45,47].

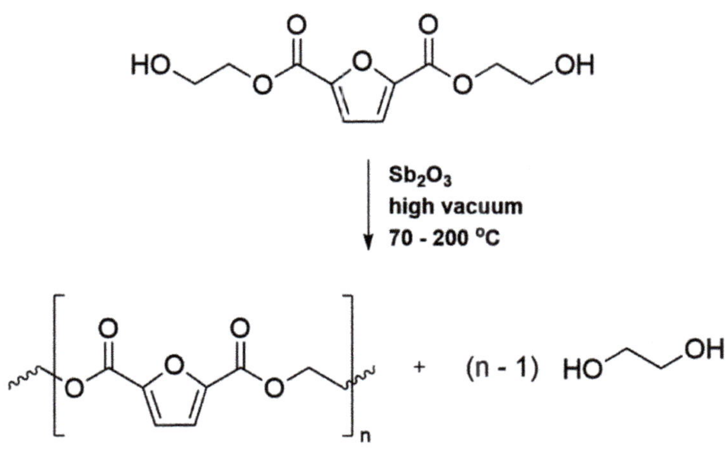

Figure 5.9 The first reported synthesis of PEF.

After almost 15 years of research and data collection, it is possible to state that, on average, both polyesters are similar in terms of thermal stability during melting in subsequent processing cycles [48], and hydrolytical stability below their Tg [40,49,50]. Some characteristics that favor PET are mostly related to mechanical performance (PEF has an increased tendency towards brittleness than PET) and color (under certain synthetic conditions, PEF is produced as a yellowish material) [40,51]. DMTA analysis indicated initial E' moduli slightly higher in the case of PEF; also, in what concerns molecular dynamics, unlike the active phenyl ring-flipping mechanism in PET, the nonlinear axis of ring rotation and ring polarity in furan greatly suppress ring-flipping, thus reducing β relaxation motions and diffusion. This was a very important consideration, that established a relationship between the observed low gas permeability of PEF and its reduced diffusion coefficient [50]. PEF has approximately twice higher water and carbon dioxide barrier properties than those of PET, at least six times higher oxygen barrier properties [52], and a higher equilibrium water uptake [53,54]. This peculiarity classifies PEF as a powerful and innovative material for food packaging applications. Table 5.2 summarizes some relevant properties of both polyesters.

Table 5.2 Comparison of some of the major characteristics of PET and PEF [55a] / MDPI / CC BY 4.0.

Property	**PET (amorphous)**	**PEF (amorphous)**
Molecular structure	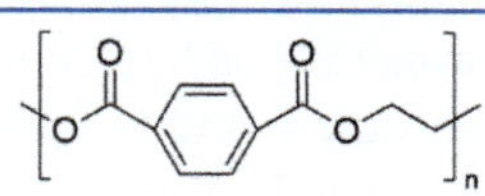	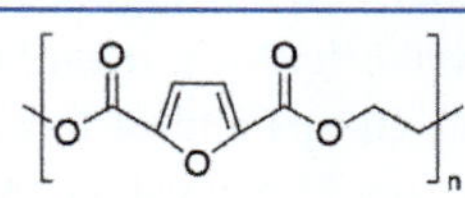
Density (amorphous)	1.36 g/cm^3	1.434 g/cm^3
Density (crystalline, calculated)	1.455 g/cm^3	1.565 g/cm^3
Melting temperature (T_m)	250–270 °C	210–230 °C
Glass transition temperature (Tg)	~76 °C	~88 °C
Crystallization time	2–3 min	20–30 min
E-modulus (ISO 527/1A, 1 mm min^{-1})	2.1–2.2 GPa	3.6 GPa
Yield strength (ISSO 527/1A, 10 mm min^{-1})	50–60 MPa	90–100 MPa
O_2 permeability (@23 °C, 65% RH)	2.5 $cm^3 \cdot mm/(m^2 \cdot 24\ h \cdot bar)$	0.23 $cm^3 \cdot mm/(m^2 \cdot 24\ h \cdot bar)$
CO_2 permeability (@23 °C, 0% RH)	23.6 $cm^3 \cdot mm/(m^2 \cdot 24\ h \cdot bar)$	1.6 $cm^3 \cdot mm/(m^2 \cdot 24\ h \cdot bar)$
H_2O permeability (@38 °C, 90% RH)	0.9 $g \cdot mm/(m^2 \cdot 24\ h)$	0.36 $g \cdot mm/(m^2 \cdot 24\ h)$

In fact, much of the success of bringing PEF towards commercialization is explained by its excellent barrier and strength properties, which allows its industrial viability despite the intrinsic higher overall production cost. The largely publicized YXY® plant-to-plastics process developed by the Dutch company Avantium for the production of commercial PEF has a solid protection of intellectual property (IP), with 57 patent families representing more than 400 patent rights and covering the full PEF value chain. Avantium has patents to produce FDCA precursors and FDCA, as well as to polymerize FDCA to PEF, whose production is now being scaled up to industrial levels. Recently, authors directly working at Avantium released a comprehensive monograph [55b] describing the path to bring FDCA and PEF to the market, including topics related to process and application development, regulatory requirements, IP protection, commercial partnerships, by-product valorization, life cycle assessment (LCA), recyclability and circular economy fit, and end-of-life considerations. At the present, Avantium is finishing the construction of the world's first FDCA Flagship Plant (situated in Delfzijl, The Netherlands), which is aimed to be operational in 2024 to produce 5000 kt/year of FDCA [55].

Concerning the routes to produce PEF, the conventional processes for polyesters synthesis, mostly involving polymerization by direct esterification or by transesterification, may sometimes correspond to time-consuming reactions conducted under high processing temperatures (around 200 °C), leading to degradation and undesired discoloration of the final polymers. Alternatively, a promising ring-opening polymerization (ROP) mechanism catalyzed by cyclic stannoxane was described in 2018 (Figure 5.10) [56]. In this case, the esterification by-products are removed *prior to the actual polymerization* by forming macrocyclic oligomers with no end-groups and of low ring strain, and the process is therefore entropically driven. A white powder of Mn higher than 40 kDa was isolated within 25 minutes, which is an excellent result. The authors mention that another advantage of using ROP in this case is the possibility of running living-like polymerizations, where polymer chains grow without termination and molecular weight can be controlled by the amount of initiator. Also, ROP is the mainstream in the industrial production of PLA [57], which indicates a viability in terms of industrial apparatus.

Besides the optimization of its synthesis, properties, and potential applications of PEF, studies have been devoted to the extension of its structure to a whole family of homologous polyesters incorporating a variety of diols [58,59], some of which included both those with linear C2 to C18 carbon chains, or instead, those with more rigid cyclic aliphatic and aromatic structures.

5.1.2.2 Other Poly(2,5-furan dicarboxylate)s

At the beginning of last decade, it became clear that the road for FDCA-based polyesters was only about to start. The demonstration of straightforward preparation methods using traditional catalyst and accessible reaction conditions,

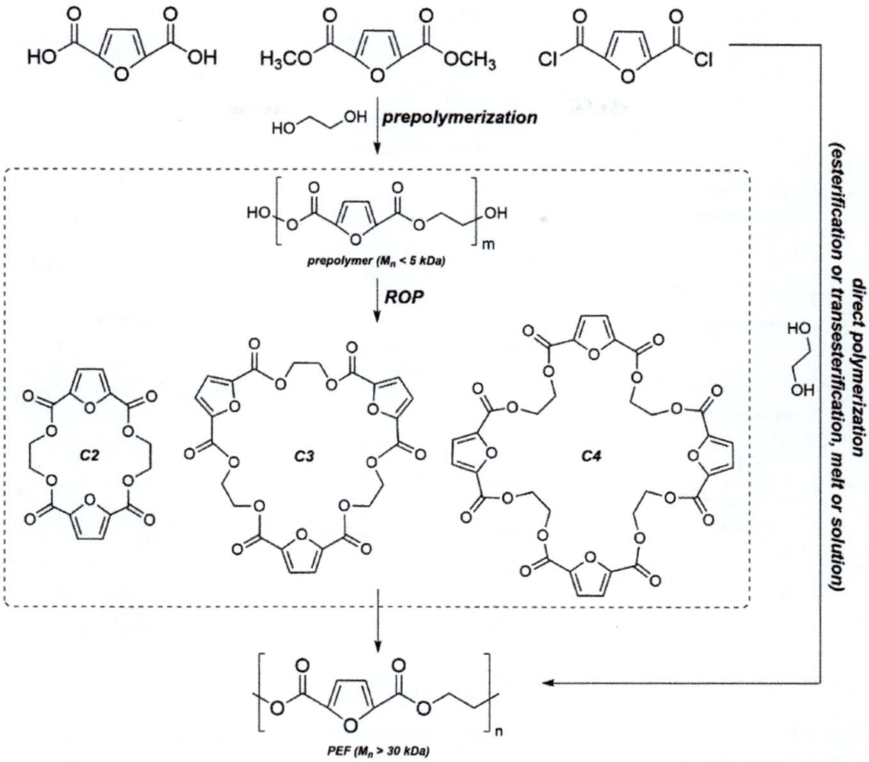

Figure 5.10 Synthesis of FDCA-based polyesters from melt polymerization, solution polymerization, and ROP [56,57].

together with the easy possibility of tunnelling the properties of the final materials by varying the specific chemical structure of the chosen diol, were powerful indications that PEF was not the only promising member of the family. Also, the use of comonomers (especially diols) that could also be obtained from renewable resources allowed the preparation of entirely renewable polyesters (Figure 5.11), which meant (and mean) an important step towards sustainable materials.

In this context, other poly(alkylenefuranoates) bearing aliphatic linear or cyclic, short-to-long chain diols, as well as aromatic diols, are continuously being reported in literature and occupy different positions in terms of relevance and product development. After PEF, the most popular ones include poly(propylene-2,5-furandicarboxylate) (PPF), poly(butylene-2,5-furandicarboxylate) (PBF), poly(hexylene-2,5-furandicarboxylate) (PHF) and poly(octylene-2,5-furandicarboxylate) (POF), with PBF occupying a privileged position for its high crystallinity and structural resemblance with poly(butylene terephthalate, PBT), a thermoplastic engineering polymer that finds applications as an insulator in the electrical and electronics

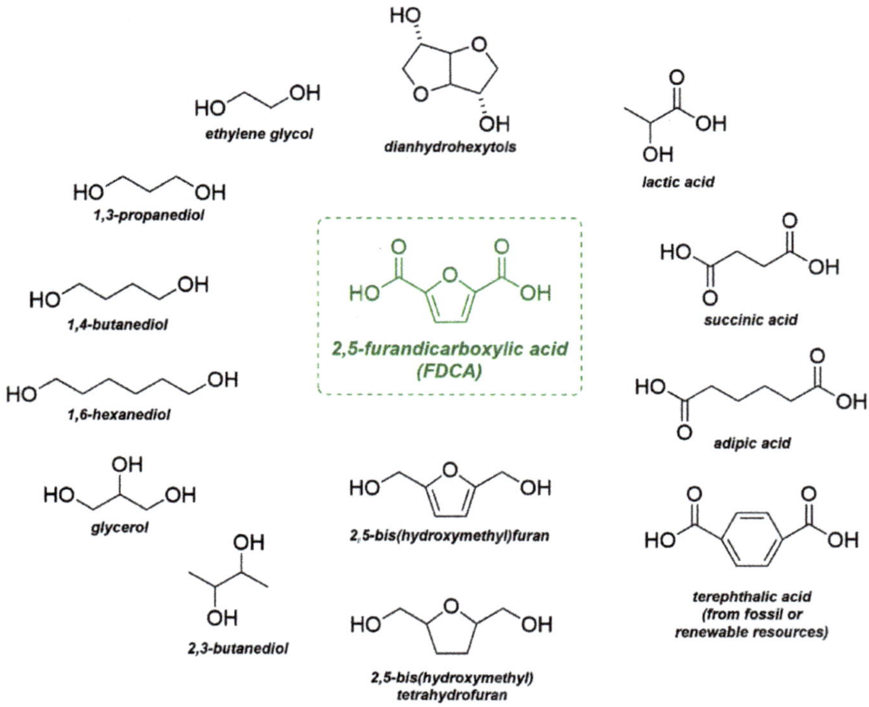

Figure 5.11 Chemical structure of several comonomers for the preparation of FDCA-based polyesters [40, 57].

industries [40]. A pioneer systematic investigation analyzing the structure/property relationship of a series of furan-aromatic polyesters based on FDCA and higher alcohols, e.g., 1,3-propanediol, 1,4-butanediol, 1,6-hexanediol, and 1,8-octanediol, and comparing them with the benzene-aromatic analogues [45], opened the way to a very broad range of possibilities. Table 5.3 summarizes some relevant thermal and mechanical properties of PEF, as well as of other poly(2,5-furandicarboxylate)s produced from diols bearing linear methylene sequences.

Researchers are now focused on improving methods and/or final properties, aiming at applying the obtained materials in specific applications. This was the case of polyesters (Mn up to 30 kDa) synthesized by two-stage melt polycondensation from the dimethyl ester of FDCA and glycols bearing three to six carbon atoms [60]. The samples were processed into thin films by compression moulding, and only the homopolymers containing the glycol subunit with an even number of -CH_2- groups (poly(butylene 2,5-furanoate) (PBF) and poly(hexamethylene 2,5-furanoate) (PHF)) were able to develop a three-dimensional ordered crystalline phase. X-ray scattering experiments using synchrotron radiation allowed the authors to deepen the understanding of important structure-property relationships

Table 5.3 Thermal and mechanical data of common linear aliphatic poly(furanoates) [47] / John Wiley & Sons.

	$M_n{\times}10^{-3}$	$M_w{\times}10^{-3}$	η_{sp} (dl g^{-1})	T_c[a] (°C)	T_g[a] (°C)	T_m[a] (°C)	T_d[b] (°C)	Young's modulus (MPa)	σ_{max} (MPa)	ε_{break} (%)
PEF	8.0	15.2	–	–	77	214	–	–	–	–
	–	–	0.47 (intrinsic)	–	89	201	326	–	–	–
	22.4	44.5	–	165	80	215	300	–	–	–
	23	–	–	156	85	170	332	–	–	–
	105	252	1.20	–	90	210	389	2070	66.7	4.2
PPF	10.1	16.3	–	–	40	171	–	–	–	–
	15	–	–	102	39	150	335	–	–	–
	21.6	27.6	–	127	50	174	295	–	–	–
	60.2	89.8	1.21	–	58	–	375	1550	68.2	46
PBF	–	–	0.45 (intrinsic)	–	36	169	329	–	–	–
	11.8	21.2	–	–	36	177	–	–	–	–
	17.8	42.3	1.41	–	31	172	373	1110	19.8	2.8
	23.2	65	–	–	39	172	–	959	31.8	1055
	60	–	–	90	31	170	338 (5%)	–	–	–
PHF	32.1	66.7	1.04	–	28	148	375	493	35.5	210
POF	20.7	47.5	0.69	–	22	149	375	340	20.3	15

a) DSC data, measured at 10 °C min^{-1}.
b) TGA data.

of the materials under study, mostly related to mechanical and barrier properties. The hypothesis previously presented by the same research group on the formation of a 2D microstructure in poly(pentamethylene furanoate) seems correct, but strongly fails in attributing to the intramolecular forces the nature of hydrogen bonds (Figure 5.12) [61]. It is important to mention that poly(furanoates) synthesized from diols with odd-number methylene groups were reported previously, although with a premature overall comprehension of the system [62]. Poly(1,12-dodecylene 2,5-furanoate) [63] and poly(1,20-eicosanediyl 2,5-furanoate) [64] were also described, with the latter being susceptible to enzymatic and hydrolytic degradation. The studies, carried out using a phosphate buffer solution under neutral pH conditions and in the presence of Porcine pancreas enzyme, showed enhanced degradability of the very-long-aliphatic furan-polyester (ca. 11% of weight loss after 21 days of exposure) when compared to PEF (see Section 7).

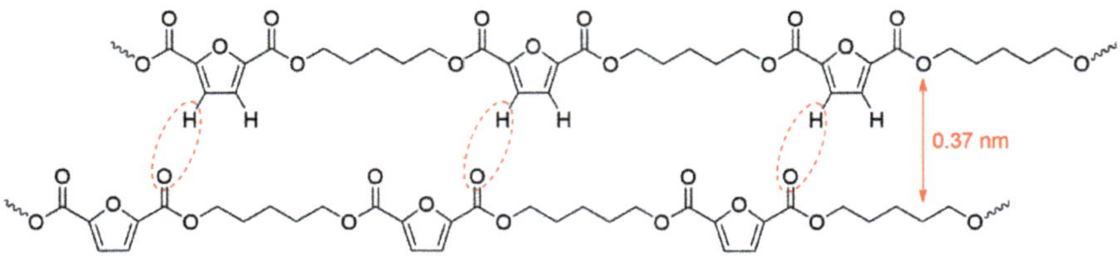

Figure 5.12 Schematic representation of the mesophase coming from interchain interactions [61].

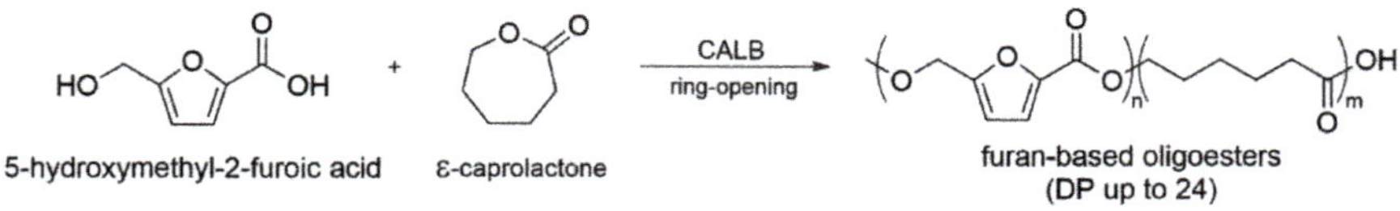

Figure 5.13 Oligoester synthesis from 5-hydroxymethyl-2-furancarboxylic acid and ε-caprolactone.

Furan-based oligoesters were synthesized by ring-opening ε-caprolactone with 5-hydroxymethyl-2-furancarboxylic acid, using immobilized *Candida antarctica* B as catalyst (Figure 5.13) [65]. The authors were able to demonstrate the formation of cyclic and linear oligomers, with polymerization degrees up to 24 and narrow molecular weight distribution (dispersity of ca. 1.1). Although the products possess a slightly different molecular structure with respect to poly(1,6-hexamethylene 2,5-furanoate), the possibility of starting the reaction from an HMF direct derivative is worth of attention. Also, the products were tested in terms of enzymatic and microbial degradation and indicated potential for long-term controlled drug delivery, which may direct the materials for high value-added applications.

The investigations considering longer-chain diols quickly shifted to branched ones at the beginning of the last decade, with a particularly increased interest for 2,3-butanediol [66–68]. A more recent contribution considered 2,5-hexandiol and 2,7-octandiol for that purpose, and the authors were able to demonstrate that the use of secondary diols led to polymers with higher glass transition temperatures (Tg) than those prepared from their primary counterparts. 2,7-octanediol, in particular, showed strong potential for tailoring the properties of its ensuing FDCA-polyesters in terms of increasing the Tg with respect to its non-branched homologue [69].

The use of diols bearing pendant groups is an efficient strategy to tune the thermal and mechanical properties of the synthesized polyesters. Alkyl and phenyl-substituted 1,3-propanediols (including 2-alkyl-1,3-propanediols and 2,2-dialkyl-1,3-propanediols with alkyl groups ranging from methyl to butyl, as well as their phenyl analogues) were reacted with the dimethyl ester of FDCA, and polyesters with Mw up to 97,100 kDa were synthesized (Figure 5.14) [70]. The results indicated higher

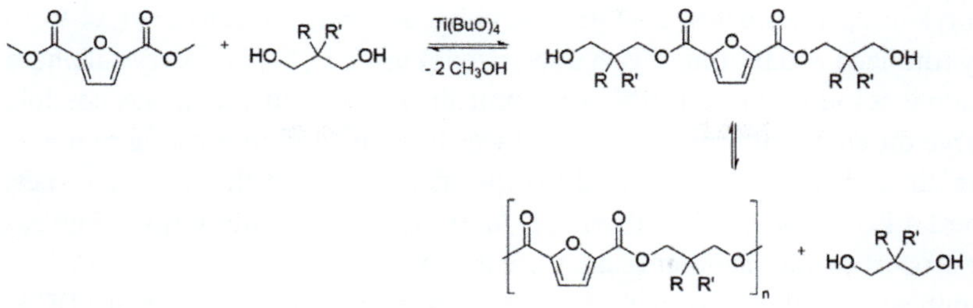

R = R' = H
R = Me, R' = H; R = Et, R' = H; R = Pr, R' = H; R = Bu, R' = H; R = Ph, R'= H
R = R' = Me; R = R' = Et; R = R' = Pr; R = R' = Bu; R = R' = Ph

Figure 5.14 Synthesis of polyesters of dimethyl-2,5-furandicarboxylate with 1,3-propanediol, 2-monosubstituted and 2,2-disubstituted 1,3-propanediols.

Tg values in the case of polymers bearing odd numbered carbon diols in the alkyl groups in comparison with those with even numbered ones. Poly(2-methyl-1,3-propylene-2,5-furanoate) was synthesized some years earlier, and thermogravimetric analysis indicated no significant weight loss up to ca. 400 °C [71].

The effect of inserting additional non-hydroxyl oxygen atoms within the polymer chain was considered for the synthesis of poly(diethylene glycol 2,5-furanoate) [72], aiming at increasing the Tg and improving the gas barrier properties of poly(pentylene 2,5-furanoate) (PPeF). The objective was indeed achieved, and a glycol-based polyester of Mn = 22.5 kDa was isolated with a Tg of ca. 20 °C higher than that prepared from 1,5-pentanediol. Non-oriented circular polyester films were prepared by hot press moulding/quenching procedures, and their gas barrier properties tested using complementary pressure-decay sorption and permeation techniques at 23 °C. The results indicated that poly(diethylene glycol 2,5-furanoate) exhibited superior O_2 barrier properties than PPeF and commercial PET tested under the same conditions, which was attributed to stronger intermolecular interactions. More recently, a similar polyester polyol, synthesized as a low-DP oligomer, was used for the preparation of polyisocyanurate rigid foams [73].

The use of isomers of FDCA was also considered to establish a structure-property relationship in poly(2,5-furan dicarboxylate)s synthesized from linear aliphatic diols. The first preliminary investigation [18c] was followed by a very thorough work conducted in 2014 [67], where polyesters based on 2,4-FDCA and 3,4-FDCA isomers with linear diols were thoroughly characterized and compared with their 2,5-FDCA analogues prepared using the same conditions. On that occasion, no clear correlation was found with respect to reactivity (yield, Mn) or selectivity (Đ, color), and thermal analysis indicated that amorphous 2,4-FDCA polyesters had Tg values similar to those of the 2,5-isomer. The potential of 2,4-FDCA motivated a continuation of the study some years later, and both poly(butylene 2,5 furanoate) (2,5-PBF) and poly(butylene 2,4 furanoate) (2,4-PBF) were synthesized and thoroughly characterized (Figure 5.15) [74]. A crystal phase formation was induced in 2,4-PBF by an annealing treatment, and it was possible to observe different patterns in crystal lattices in each case. Interesting to notice that the different isomerism affected the thermal stability of the final materials, with the 2,4-isomer being more thermally inert than the 2,5- counterpart. The gas barrier properties, on the other hand, were much better for 2,5-PBF.

A significant number of contributions can also be found in the case of FDCA-based polyesters bearing cyclic diols, although in large minority with respect to those tackled so far. The approach gains relevance when one considers the possibility of incorporating dianhydrohexytols (Figure 5.11) within the polymer backbone, for the complementarity of the building blocks in terms of renewable origin. However, the rigidity impaired by the bicyclic structure of such diols, together with the inherent rigid nature of the furan ring in FDCA, often yield brittle final materials of low DP, and the simultaneous use of both monomers becomes much more frequent in the case of copolyesters, where the final composition of the polymer may be tailored considering the use of other comonomers with increased chain flexibility. A smart strategy to overcome such limitation was published in a study [75], where isoidide-2,5-dimethanol was synthesized and polymerized with FDCA through melt polycondensation followed by solid-state post-condensation. This was a pioneer investigation on the preparation of semi-aromatic homopolyesters of

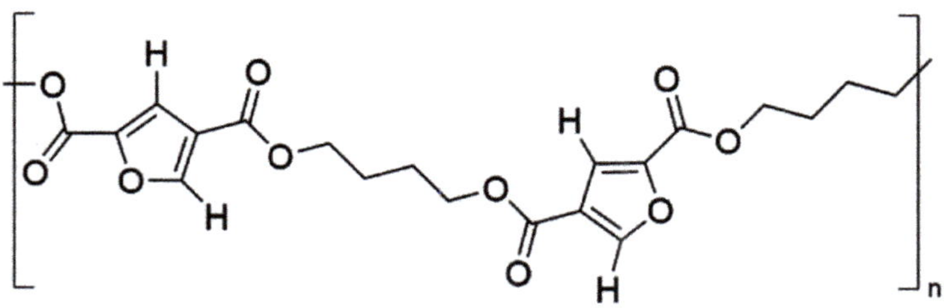

poly(butylene 2,4-furanoate) (2,4-PBF)

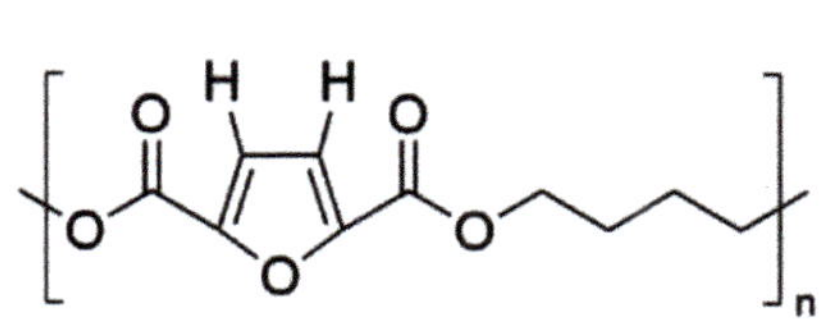

poly(butylene 2,5-furanoate) (2,5-PBF)

Figure 5.15 Chemical structures of 2,4-PBF and 2,5-PBF, considering two repeating units with head-to-tail and tail-to-head orientations of a furan moiety for 2,4-PBF [74].

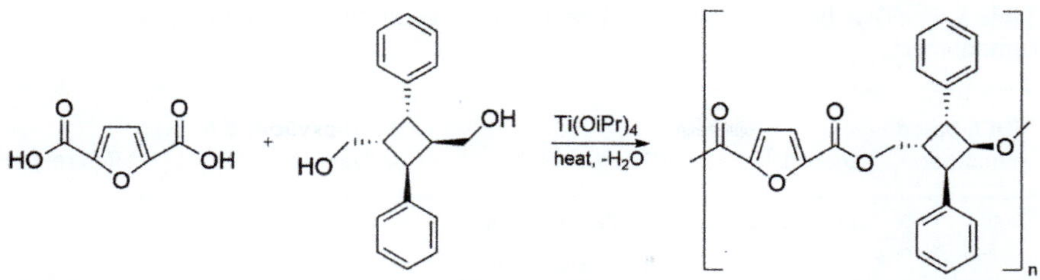

Figure 5.16 Synthesis of a polyester from FDCA and trans-1,3-cyclobutane dimethanol, prepared from cinnamic acid via UV-mediated photocyclization [76].

high Mn (ca. 30 kDa) via an industrially relevant procedure using isohexide derivatives. More recently, FDCA was polymerized with a semi-rigid cyclobutanediol monomer (Figure 5.16) in a two-step procedure. A pre-polymerization was carried out at 160 °C for 12 h, followed by further polycondensation at 210–215 °C for 2 h under reduced pressure, in the presence of titanium (IV) isopropoxide as catalyst (1.25 mol %). A polyester of Mn = 3600 and Mw = 6900, Tg of 109 °C and high thermal stability was isolated, indicating good potential to replace BPA-based polyesters [76].

Other existing strategies to adjust the properties of FDCA-based polyesters are mostly related to copolymerization, and a series of diols and/or diacids have been employed as comonomers for this purpose. Table 5.4 summarizes some relevant contributions to the field published within the last few years.

The deliberate choice of examples used to build Table 5.4 was obviously not made with the intention of covering the whole family of FDCA-based copolyesters, neither to indicate those that have increased potential or importance than others that were neglected here. It becomes clear, by observing the broad spectrum of possible combinations, that a virtually endless source of original biobased materials may be engineered to achieve improved properties, such as thermal resistance [121] and gas barrier for different gases [105], and to address specific applications, ranging from 3D printing [109] to liquid crystals [122]. The idea is even wider when one considers the utilization of FDCA isomers, which was the case of highly transparent films of poly(ethylene terephthalate)-co-(ethylene 2,4-furandicarboxylate), whose properties were tailored by varying the furan unit content [123]. Also, the simultaneous addition of other comonomers such as diamines for the synthesis of furan-based poly(ester amides) [124], may favor increased mechanical and thermal properties with respect to the polyesters *per se*.

The large development related to polyesters based on monomers bearing a single furan core also includes those derived from bis(hydroxymethyl)furan (BHMF) (Figure 5.7b). The preparation of poly(2,5-furandimethylene succinate) was first reported in 2013 [125] for the reaction of BHMF with succinic acid (1 equiv. mol) in the presence of *N,N*-dimethyl-4-aminopyridine and

Table 5.4 FDCA-based copolyesters bearing linear, cyclic, aliphatic, and aromatic comonomers.

FDCA-based monomer	Diol #1	Diol #2	Dicarboxylic acid (or its derivative)	Reference
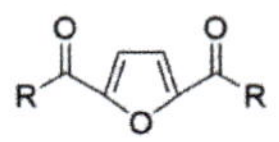	Ethylene glycol	1,5-Pentane diol	–	[77]
		1,6-Hexane diol	–	[78]
R = OH, O-alkyl, Cl		–	Lactic acid	[48]
		Diethylene glycol	–	[79]
		Diethylene glycol	Dimethyl adipate	[80]
		–	Ethyl vanillate	[81]
		1,4-Cyclohexane dimethanol	2,3,5,6-Tetrafluoro-1,4-benzenedimethanol	[82]
		–	Terephthalic acid	[83]
		Polycarbonate diol*	–	[84]
		Poly(tetramethylene oxide) diol	–	[85]
	1,3-Propane diol	1,4-Cyclohexane dimethanol	–	[86]
		Fatty acid dimer diol	–	[87]
		1,2-Propane diol	Itaconic acid	[88]
		–	Succinic acid and cumaric acid	[89]
		–	Itaconic acid and succinic acid	[90]
		Isosorbide	Fumaric acid	[91]
		Isosorbide	Succinic acid	[92]
		Isomannide	–	[93]
		Isosorbide	–	[94]

Table 5.4 (Continued)

FDCA-based monomer	Diol #1	Diol #2	Dicarboxylic acid (or its derivative)	Reference
	1,4-Butane diol	–	Isosorbide carbonate**	[95]
		1,4-Cyclohexane dimethanol	–	[96]
		–	Tricyclic diacids and diesters***	[97]
		Fatty acid dimer diol	–	[98]
		Poly(ethylene oxide)	–	[99]
		–	Dimethyl terephthalate	[100]
			Terephthalic acid and sebacic acid	[101]
		–	2,6-Naphthalene dicarboxylic acid	[102]
		–	2,5-Thiophene dicarboxylic acid	[103]
		–	Dimethyl succinate	[104]
		N-(2-Hydroxyethyl)-4-hydroxybutyramide	–	[105]
		–	Diglycolic acid	[106]
	2,3-Butane diol	Various polyfunctional alcohols	–	[107]
	1,6-Hexane diol	Isosorbide	–	[108]
		1,2-Pentane diol	–	[109]
		Triethylene glycol	–	[110]
		Poly(ethylene glycol)	–	[111]
		Poly(tetrahydro furan)	–	[112]
		–	Sebacic acid	[113]
		–	Itaconic acid	[114]
		–	2-Carboxyethyl (phenyl) phosphinic acid	[115]

(Continued)

Table 5.4 (Continued)

FDCA-based monomer	Diol #1	Diol #2	Dicarboxylic acid (or its derivative)	Reference
	1,8-Octane diol	–	Terephthalic acid	[116]
	1,10-Decane diol	Amidodiol****		[117]
	Various aliphatic diols	–	Terephthalic acid	[118]
	Isosorbide	–	1,12-Dodecanedioic acid	[119]
		–	1,9-Nonanedioic acid (azelaic acid)	[120]
	1,4-Cyclohexane dimethanol	Poly(tetramethylene glycol)	–	[121]
	1,1′-Biphenyl-4,4′-diol	–	2,6-Naphthalene dicarboxylic acid and sebacic acid	[122]

Synthesized from: * 1,5-Pentanediol, 1,6-hexanediol and dimethyl carbonate; ** Isosorbide and dimethyl carbonate; *** Furan and maleic anhydride, via Diels–Alder coupling; **** Via ring opening reaction of ε-caprolactone with 1,4-aminobutane.

N,N′-diisopropylcarbodiimide at room temperature. The semi-crystalline polyester (Tg = 15 °C, Tm = 79 °C) was then crosslinked using 1,8-bis(maleimido)-triethylene glycol, via the thermo-reversible Diels–Alder reaction, which will be thoroughly addressed in this book in Section 6. The exact same approach was published afterwards by the same research group, however including 1,3-propanediol or 1,4-butanediol as comonomers [126,127]. Poly(2,5-furandimethylene succinate) (Mn = 4620 Da) was synthesized from BHMF and succinic acid using the same catalytic system, and then applied within a multiblock system that included poly(butylene succinate) as the other block and 1,6-hexamethylene diisocyanate as chain extender [128].

In another vein, the application of a *N*-heterocyclic carbene was tested for the polycondensation of BHMF with benzene-1,4-dialdehyde and furan-2,5-dialdehyde under oxidative conditions [129]. The materials were isolated in high yields and with molecular weights (Mn) of 3.1 kDa and 7.8 kDa, respectively. Among all eleven polyesters that were investigated in this study, the more stable was the one derived from BHMF and terephthalaldehyde, which, despite its relatively low molecular weight, was found to be a semicrystalline polymer with a Tg higher than room temperature (60 °C) and Tm of ca. 180 °C, thus indicating that it is a promising candidate for the preparation of transparent thermoplastic films. The authors also considered the polymerization reactions starting from ethyleneglycol and furan-2,5-dialdehyde or 5,5′-[oxybis(methylene)]bis[2-furancarboxaldehyde].

The production of BHMF from HMF is an established process, using both homogeneous and heterogeneous catalysts, but HMF from renewable resources is often contaminated with colored polymeric impurities called humins, and the search for alternative pathways is desirable in order to obtain high quality colorless final materials, mainly for packaging applications [130]. That was the main motivation for the design of a multicatalytic route to produce 2,5-bis(hydroxymethyl)furan and the corresponding (co)polymers starting from FDCA. The authors demonstrated a successful one-pot procedure to produce BHMF with a high yield and selectivity from the FDCA dimethyl ester, and the polymerization reaction of BHMF with adipic or pimelic acid was carried out under mild conditions, using di-*t*-butyl dicarbonate (Boc_2O) as a coupling agent and a Lewis acid catalyst. Polyesters with molecular weights (Mn) up to 6410 Da were obtained without the need of isolating and purifying BHMF.

Other interesting initiatives to polymerize BHMF are based on enzymatic polycondensation [51,58,131]. BHMF was oligomerized with various diacid ethyl esters by *Candida antarctica* Lipase B (CALB) in diphenyl ether [132], and a further step was conducted under vacuum as a tentative method to obtain high molecular weight polyesters, but polymers with Mn values only up to 2400 Da were isolated, which was explained by the occurrence of etherification under the conditions tested. A complementary investigation allowed the authors to observe the distinct activity of the enzyme towards different building blocks [133] and the copolymerization of BHMF with FDCA dimethyl ester and aliphatic diols or diacid ethyl esters of varying chain lengths yielded semi-crystalline polyesters with DPs up to 122 [134]. Another very interesting study described the utilization of both 2,5-BHMF and 3,4-BHMF for the enzyme-catalyzed synthesis of polyesters [135]. The polycondensation reactions gave isolated yields higher than 65%, except in the case of 3,4-BHMF (ca. 50% yield). The 2,5-BHMF-based polyesters were isolated with Mn values up to 3100 Da, depending on the aliphatic diester used.

5.1.3 Polyesters from Furanacrylic Monomers

In the first investigations of furan-based polyesters [17], the hydroxyacid depicted in Figure 5.17a was tested with no success as an AB polycondensation monomer, with only linear and cyclic oligomers being formed. However, it was proven to be a very useful intermediate for furanacrylic monomers (Figure 5.17b and c), and a pioneer study described the preparation of furan-based unsaturated polyesters by transesterification polycondensation of 2-hydroxymethyl-5-furanacrylic acid ethyl ester (Figure 5.18) [136].

The purpose of this study was to prepare a photocrosslinkable polymer to be used as a photoresist in printing plates, but low Tg, flexible film-forming materials were also described [136]. Thanks to the presence of the external unsaturation in the furan comonomer, irradiation of the films with UV light induced the photocrosslinking required for image stabilization, as depicted in Figure 5.19.

The quite simple practical aspect of such photocyclization is desirable in terms of polymer synthesis. The ring strain energy of cyclobutane is about 20.1 and 26.2 kcal mol^{-1} higher than that of cyclopentane and cyclohexane, respectively [137,138], which may be a positive aspect depending on the specific application. Stimulating work has significantly contributed to the development of furan-based polyesters using such photocyclization as a key tool. Such strategy was used in order to prepare a bifunctional furanacrylic monomer by reacting 1,5-pentanediol

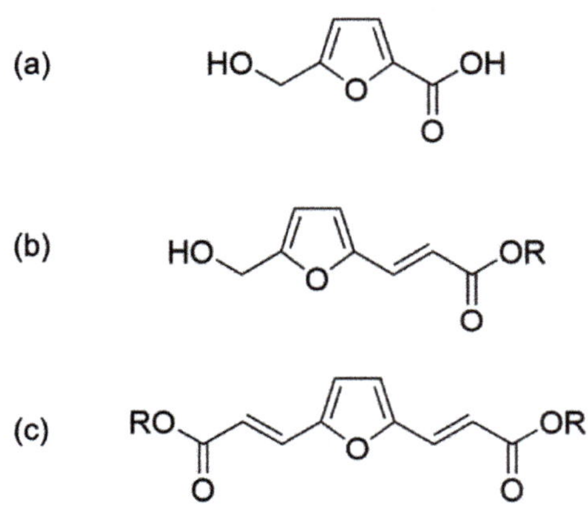

Figure 5.17 Molecular structures of (a) 5-hydroxymethyl furoic acid and (b,c) furanacrylic monomers.

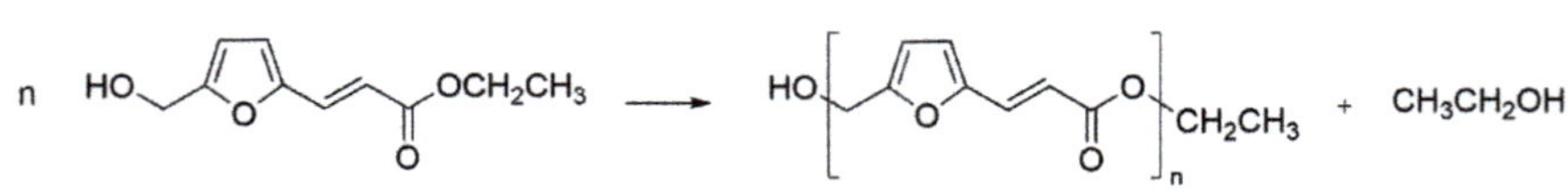

Figure 5.18 Transpolycondensation reaction of unsaturated furan difunctional monomer.

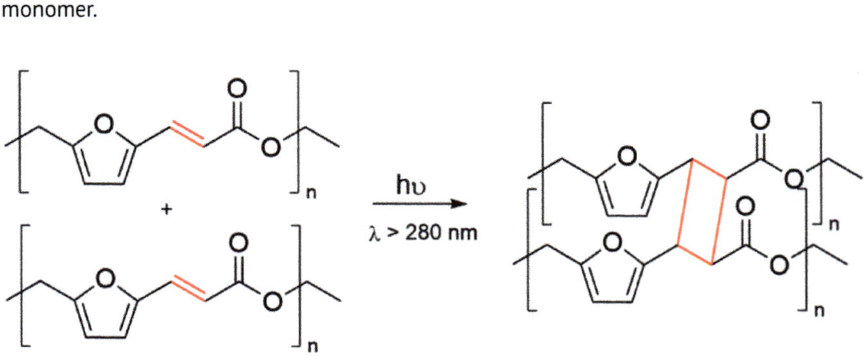

Figure 5.19 Photocrosslinking through the external unsaturation of the furan comonomer.

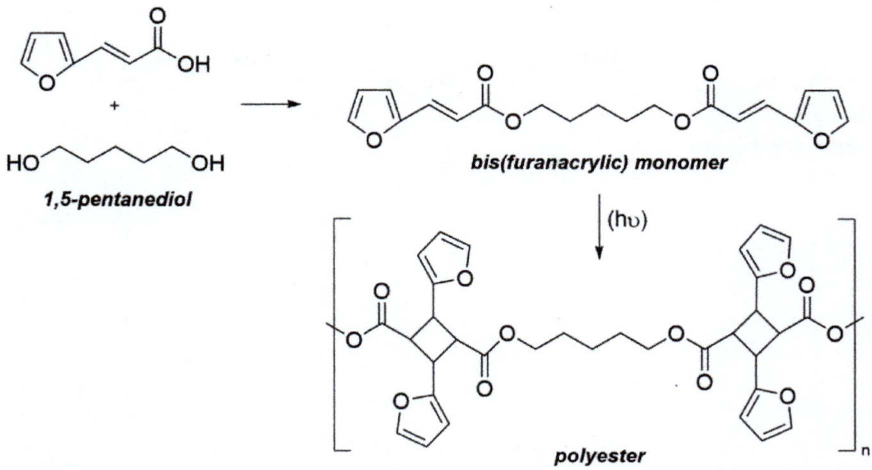

Figure 5.20 Synthesis and photopolymerization of a bifunctional furanacrylic monomer.

with 2 equiv. of 2-furanacrylic acid [139]. The monomer was then submitted to a solvent-free polymerization reaction using sunlight, as represented in Figure 5.20, which was surprisingly insoluble in common organic solvents (e.g., acetone, acetonitrile, ethyl acetate, toluene, DMF and DMSO).

The relevance of the study goes far beyond the synthesis and characterization of the furan polyester itself, as the authors present an important discussion on the stereospecificity of the [2+2] photochemical coupling, as a result of X-ray diffraction analysis. As a matter of fact, stereoregularity directly affects the performance of the polymeric materials, and stereoregular polymers often perform better than the corresponding non-stereoregular ones [140,141]. Because a topochemical reaction normally proceeds with minimum movement of atoms, the authors observed the formation of only one of the five possible stereoisomers of the [2 + 2] head-to-tail coupling on the polymer backbone.

The same research group then shifted to monomeric systems and focused on a deeper understanding of crystallographic arrays, as well as of the cyclobutene core stability in monomers and polymers [138,141,142]. Trans-cinnamic acid was submitted to photocyclization for the synthesis of *trans*-2,4-diphenylcyclobutane-1,3-dicarboxylic acid (Figure 5.21) as a unique semi-flexible/semi-rigid building block for materials design, as the presence of a core cyclobutene limits conformational freedom [138]. The photocyclization involved head-to-tail packing because it is lower in energy. The authors reacted the diacid monomer with ethylene glycol, 1,3-propanediol, 1,4-butanediol, 1,5-petanediol, and 1,6-hexanediol, and the semi-crystalline polyesters exhibited thermal, chemical, and photochemical stabilities comparable to those of PET.

The initiative then advanced to furanacrylic monomers (Figure 5.22), and in contrast to the major product of the cyclization of cinnamic acid, the first

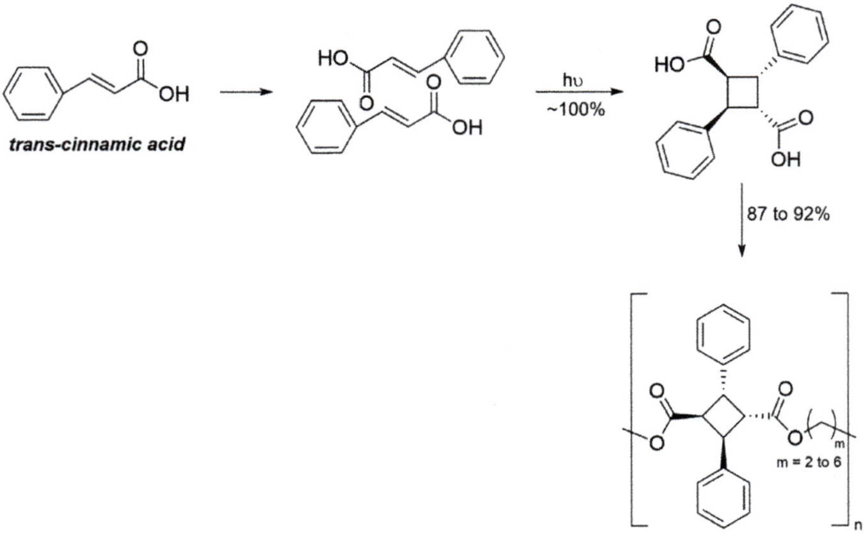

Figure 5.21 Synthesis and ensuing polymerization of trans-2,4-diphenylcyclobutane-1,3-dicarboxylic acid [138].

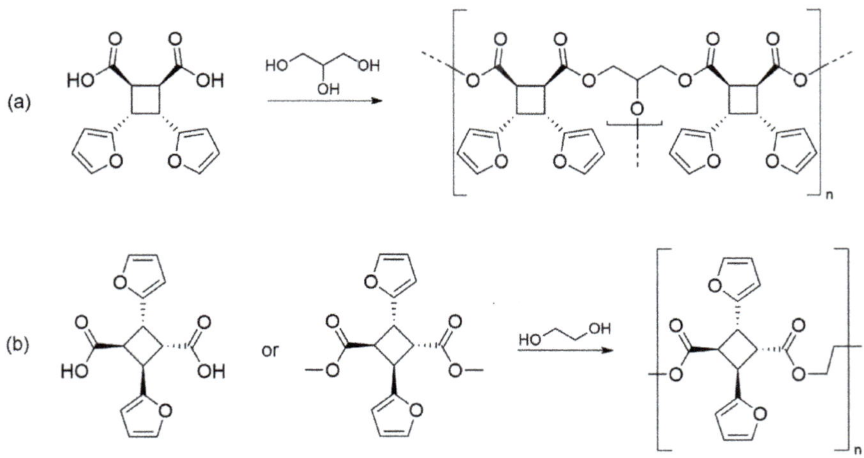

Figure 5.22 Synthesis of (a) crosslinked polyesters from *cis*-3,4-di(furan-2-yl) cyclobutane-1,2-dicarboxylic acid and glycerol and of (b) linear polyesters from *trans*-3,4-di(furan-2-yl)cyclobutane-1,2-dicarboxylic acid and ethylene glycol [141,142].

investigation on the photocyclization of furanacrylic acid resulted in a *head-to-head* packing of the molecules in the solid state [141]. The product (*cis*-3,4-di(furan-2-yl)cyclobutane-1,2-dicarboxylic acid) was then reacted with ethanol or glycerol by activating the diacid with oxalyl chloride in the first step, and the products were generated by the esterification reaction in the presence of triethylamine as a base. The reaction with glycerol yielded an insoluble crosslinked polymer that was isolated as a fragile film at room temperature, with a Tg of 50 °C. More recently, the authors reported a successful synthetic route to prepare *trans*-3,4-di(furan-2-yl) cyclobutane-1,2-dicarboxylic acid, which was based on a first esterification of furanacrylic acid with ethanol [142]. The hypothesis that ethyl 2-furanacrylate could offer the desired packing in the solid state, because the strong intermolecular hydrogen bonds between the carboxylic acid groups were removed, was confirmed and both acid and ester forms of the furan cyclobutane were polymerized with ethylene glycol. Gel permeation chromatography analysis indicated the formation of mixtures of polymers (Mn: ca. 22 kDa) and oligomers (Mn: ca. 2 kDa).

It is worth emphasizing that the discussion on the stereochemistry of the photochemically synthesized cyclobutanes is based on robust spectroscopic data, which opens the way for further investigations. *Cis*-3,4-di(furan-2-yl)cyclobutane-1,2-dicarboxylic acid was recently used to prepare polyester thermosets by ring-opening the oxirane groups of epoxidized linseed oil [143]. The authors closely inspected the polymerization kinetics and chemorheology of this system, and this elucidated important mechanistical aspects. The esterification reaction of furanacrylic acid and diethylene glycol or trimethylolpropane yielded a bi- and a trifunctional furanacrylate, which were used to prepare linear and crosslinked polyesters, respectively, by photopolymerization [144]. Another approach describes the preparation of original furanacrylates bearing terminal double bonds that were used as polymerization sites via thiol-ene reaction with tris- and tetrafunctional thiols (Figure 5.23) [145]. According to the aforementioned investigations, although the authors present here the structure of *trans*-3,4-di(furan-2-yl)cyclobutane-1,2-dicarboxylic acid, it is most likely that the *cis*-isomer was synthesized. Interestingly, in the case of the furan cyclobutane, both monomer and polymer were synthesized

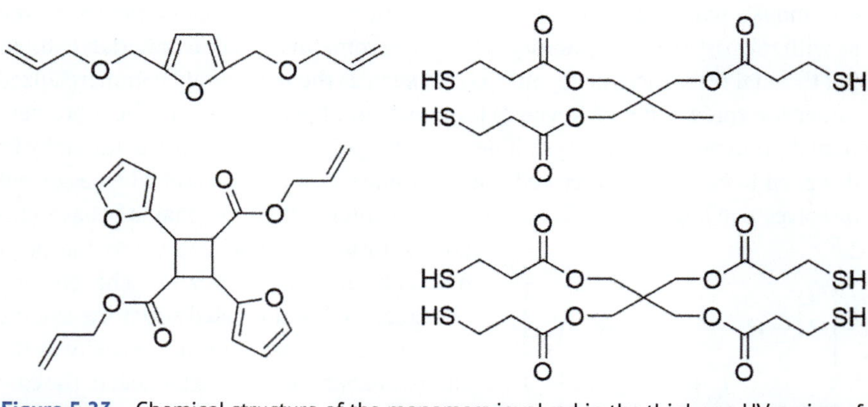

Figure 5.23 Chemical structure of the monomers involved in the thiol-ene UV-curing of furan polyesters.

under UV radiation, which is desirable considering the milder conditions needed for high conversions. The thiol-ene polymerization reactions were investigated by means of real-time FT-IR, photo-DSC and photorheology, and the final poly(thioesters) were characterized by DMTA and tensile tests, demonstrating a potential for coating applications.

5.2 Polyamides

Polyamides are a broad class of polymeric materials bearing amide linkages within their backbones (Figure 5.24), ranging from wholly linear aliphatic polyamides – that are commonly known as *nylons* – and mixed polyamides prepared from aromatic diacids and aliphatic diamines (or vice versa), to fully aromatic polyaramids. Polyamides may be produced through the condensation of diamines with dicarboxylic acids (or their analogous esters and diacyl chlorides), or through ring opening polymerizations of lactams, which is the preferable route for the industrial production of nylons [1a].

The hydrogen bonds between -C=O and -NH groups of neighboring chains largely influence the overall intermolecular conformation, leading to crystal structures in the form of sheets bearing macromolecules packed in planar *zig-zag* conformations. This explains the higher Tm values of polyamides (which are hydrogen bonded), when compared to polyolefins of the same degree of polymerization; also, the melting points of polyamides decrease with increasing lengths of hydrocarbon sequences between amide groups. Typically, nylon polymers display notable impact strength, resilience, flexibility, and resistance to abrasion, but the obvious peculiarities of the numerous commercially available nylons lie in the length of the aliphatic sections situated between the amide linkages [1].

The properties of mixed polyamides and polyaramids, on the other hand, vary enormously with respect to the molecular nature of the starting monomers, as well as with microstructure regularity, which may sometimes lead to materials exhibiting thermal resistances of beyond 500 °C, such as the polyaramid commercialized under the trade name of Kevlar [1]. In any case, however, due to the aforementioned proneness to form crystalline domains, polyamides microstructure may be designed to be further processed into fiber materials. In this case, when a solvent dissolves a nylon (flexible nonaromatic polyamide), then the chains behave randomly, forming entanglements as the polymer concentration increases. The ensuing fiber retains this entangled structure, leading to a decrease in modulus and tenacity as the chains are only partially extended; in the case of solutions of rigid rod-like aramids, however, random coil structures are absent, and

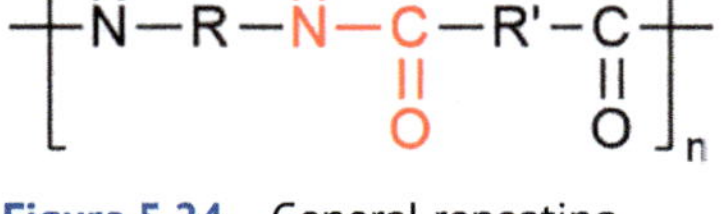

Figure 5.24 General repeating unity of polyamides.

the chains pack together in parallel bundles as the polymer concentration increases. Spinning from these solutions results in highly oriented, fully extended chains that can easily crystallize, leading to fibers with a high modulus. This is due to the shearing forces that align the bundles in the direction of the applied force during spinning [1c].

The excellent structural and thermal properties of polyamides, and the versatility impaired by the possibility of tunneling them for specific uses in engineering, have allowed an impressive widening in commercial applications since 1940, when nylon stockings were first offered for sale in New York City. At the present moment, clothing, hosiery, toothbrushes, sutures, catheters, and prosthetic films are only a few examples of the numerous polyamide-based products for daily life. Researchers have conducted extensive studies to improve final properties, such as incorporating specific functional groups in the polymer backbone, extending the chain length, and creating polymer blends and networks [146].

The production of polyamides from renewable resources dates back far before the current trend of other polymer commodities, with examples including Arkema's Rilsan polyamide developed in the 1950s from castor oil and, since then, used in coating and adhesives applications. Within this context, furan-based polyamides occupy an important position for their well-demonstrated potential to be a source of high-performance materials. The sustainable character and the pseudo-aromatic nature of the furan ring make it an attractive alternative to the design of mixed polyamides and polyaramids as it is expected to impair a more rigid character than a cyclohexyl ring, for example, thus producing properties superior to that of aliphatic and cycloaliphatic analogues [146]. In this case, considering the wide range of possibilities of using furan-based building blocks that contain amine and/or carboxylic acid functional groups (or their analogous esters and acyl chlorides), there are virtually no limits for the design and expansion of a very promising class of original sustainable materials. Indeed, this field has been thoroughly reviewed recently [147]. For the sake of clarity, the choice of interesting contributions made to the field since the beginning of the 1990s will be separated here in subsections that take into account the chemical nature of the furanic monomer, or its incorporation as pendant structures that allows the production of reversible crosslinked materials.

5.2.1 Polyamides from Bisfuranic Dicarboxylic Acids (or Their Analogous Esters and Acyl Chlorides)

Typical step-growth reactions are normally carried out in bulk, in temperatures ranging from 150 to 250 °C to accelerate the reactions and to promote the removal of condensation products of a low molar weight. However, in many cases, such harsh condition may lead to side-reactions, and room-temperature reactions using highly energetic reactants may be preferable. The Schotten–Baumann reaction, that allows the formation of amides from amines and acid chlorides, may be conducted in a two-phase system in which the diamine is soluble in both phases and diffuses across the interface into the organic layer, where the polymerization takes place. Continuous polymer production can be achieved by either withdrawing the film formed at the interface to enable the reactants to diffuse continuously, or by vigorously stirring the system to maintain a constantly changing interface, increasing the available surface area for the reaction. Since both methods are controlled by diffusion, the requirement for precise stoichiometric control is eliminated [1c].

The interfacial polycondensation of 2,2-bis(5-chloroformyl 2-furyl)propane (Figure 5.25) with hexamethylene diamine opened a consistent series of studies on the synthesis and properties assessment of furan-based polyamides [148]. The choice of such bisfuranic monomer was explained by its easiness of preparation from commercial furoates and by the good results obtained when it was applied in the synthesis of polyesters. The investigation considered different procedures, and the interfacial polymerization was demonstrated to be the most appropriate one for furanic-aliphatic polyamides, which was based on reaction yields, and on the molecular weights of final materials (assessed by viscosimetry). Two other publications that followed this initial investigation were reported in the early 2000s and consisted in the preparation of a homologous series of polyamides based exclusively on the acyl chloride version of the same furan monomer via interfacial polycondensation, but using a wide range of other aliphatic and aromatic diamines as comonomers [149,150].

More recently, 2,2'-furan dimer 5,5'-dicarbonyl chloride was reacted with various 1,ω-diamines (Figure 5.26) via solution or interfacial polycondensation to produce the corresponding polyamides with Tm values higher than those of the analogous polyamides bearing a single furan ring core [151].

Although these studies are highly relevant in terms of originality, the real turning point for furan-based polyamides was observed after the consolidation of

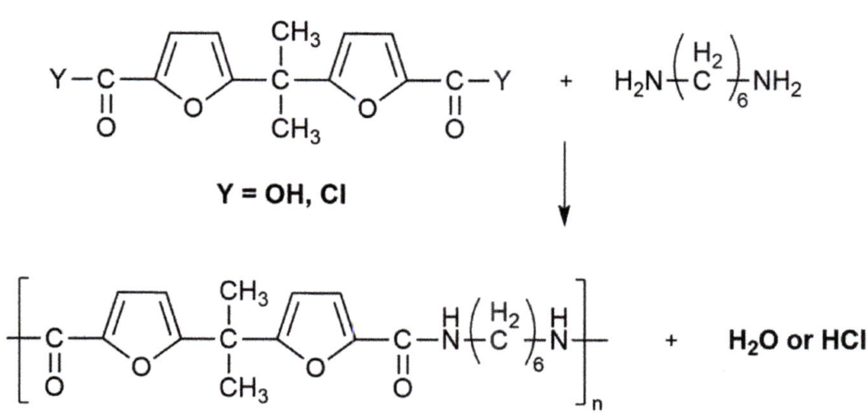

Figure 5.25 Bisfuranic dicarboxylic acid and diacyl chloride used for the preparation of furanic polyamides.

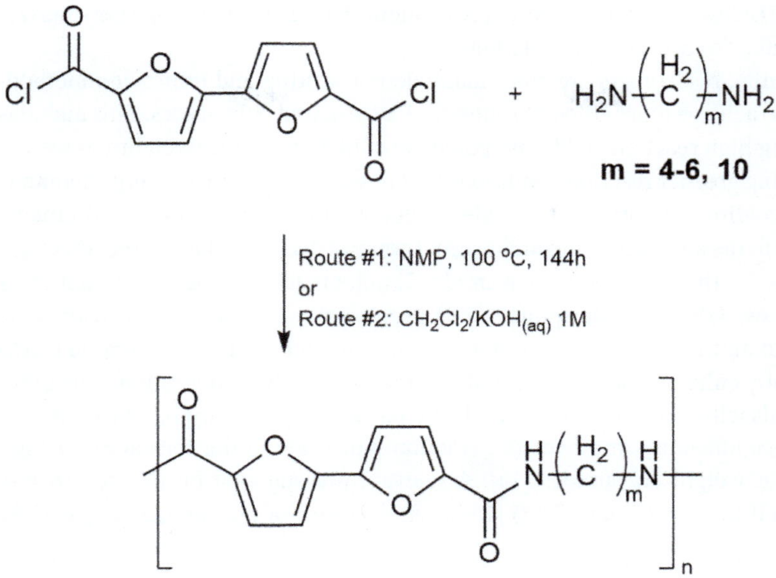

Figure 5.26 Polymerization of difuran diacyl chloride with 1, ω -diamines by solution or interfacial polycondensation.

furan dicarboxylic acid (FDCA) as a key building block for polymer synthesis. It was then obvious to utilize it for preparing polyamides, and before the strategy was demonstrated experimentally, a computational study was published to estimate thermal and mechanical properties of such furan polyamides in comparison to fossil-based nylons [152].

5.2.2 Polyamides from Furan Dicarboxylic Acid (FDCA, or Their Analogous Esters and Acyl Chlorides)

A monograph of Loos' research group – already cited within the context of polyesters – also brings a very informative contextualization of what is now the most promising routes for furan-based polyamides, i.e., those starting from furan dicarboxylic acid (FDCA), dimethyl 2,5-furandicarboxylate, and 2,5-furandicarbonyl dichloride (Figure 5.7a) [59b]. The authors gathered relevant historical events, mentioning that it was in the early 1960 that the first test of heating a hexamethylene diammonium FDCA salt was conducted with the aim of producing polyamides, which failed due to strong decarboxylation. The determination of the exact temperature of decarboxylation of FDCA (195 °C) allowed the experimental conditions to improve, and FDCA, dimethyl 2,5-furandicarboxylate, and 2,5-furandicarbonyl dichloride were successfully polymerized with diamines using different reaction conditions and metal catalysts. The companies Avantium and Solvay later reported an improved method to produce higher molecular weight FDCA-based polyamides, in two patents based on a two-stage polymerization strategy to prevent side reactions.

Currently, researchers are continually demonstrating and improving methodologies to achieve more efficient routes for FDCA-based polyamides. This includes achieving high reaction yields and polymers of high molecular weights, as well as developing greener reaction conditions that possess low intrinsic energy demands while avoiding the formation of side products. Considering these requirements and due to the sometimes-inevitable occurrence of decarboxylation, the direct use of FDCA for the synthesis of polyamides is notably less frequent than that of its derivatives. Triphenyl phosphite (TPP) promotes reactions between carboxylic acids and amines, and the addition of metal salts such as lithium chloride (LiCl) can greatly enhance the process, enabling the direct polycondensation of dicarboxylic acids with diamines. Using a solvent, particularly *N*-methyl-2-pyrrolidinone (NMP), significantly improves the reaction and leads to the formation of high molecular weight polyamides [1a]. A relevant investigation on this regard was conducted in 2016 (Figure 5.27) [153], with a series of polyamides prepared by

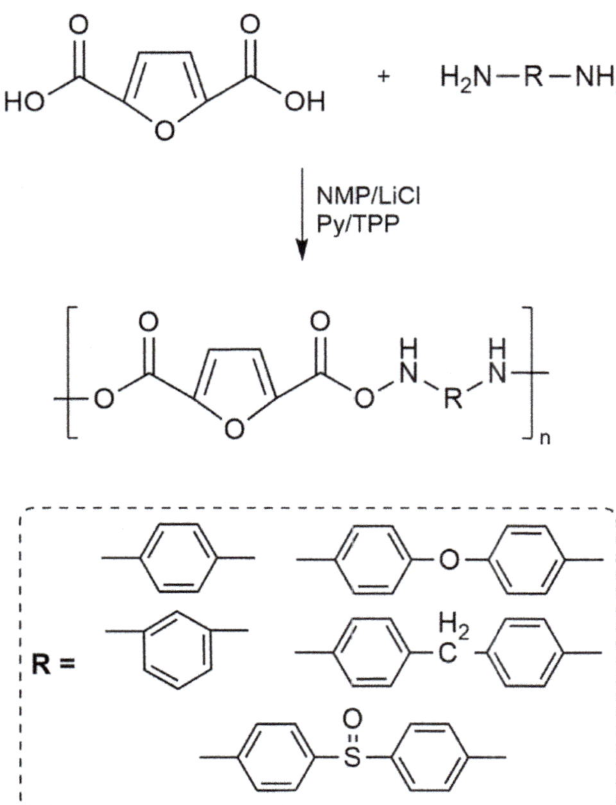

Figure 5.27 Aromatic polyamides prepared directly from FDCA and different diamines.

dissolving FDCA and LiCl in NMP, which was followed by the addition of equimolar amounts of aromatic diamines together with pyridine and TPP. The mixture was gradually heated to 130 °C and maintained at that temperature for 4 h under nitrogen atmosphere. The polyamides were isolated via precipitation in acetone and displayed Mn values ranging from 39.3 to 83.8 kDa. A different strategy was described a few months later, considering an existing industrial two-step process for the production of polyamides involving prepolymerization followed by solid-state polymerization [154]. In this case, the authors replaced varying amounts of terephthalic acid with FDCA and analyzed the relative properties of the final materials.

Shifting to FDCA derivatives may be useful to produce polyamides under mild reaction conditions. La Scala and coworkers conducted a thorough investigation on the synthesis of furan polyamides (Mn = 23.7–62 kDa) using interfacial polymerization, starting from 2,5-furandicarbonyl dichloride, aromatic diamines, and varying chain-length aliphatic diamines [146]. Aiming at contributing to the previous efforts made to improve the solubility of aromatic polyamides, 2,5-furandicarbonyl dichloride was reacted with *m*-phenylenediamine or *p*-phenylenediamine and *m*-phthaloyl chloride and/or terephthaloyl chloride in the presence of *N*-methyl-2-pyrrolidone (NMP), and polyamides of increased solubility with respect to their individual precursors were produced [155]. More recently, 2,5-furandicarbonyl dichloride was reacted with 3,4'-diaminodiphenyl ether in N,N-dimethylacetamide/LiCl at −5 °C [156]. An aromatic polyamide fiber was obtained by dry-jet wet spinning, and its structure and properties were investigated.

The routes for FDCA-based polyamides starting from dimethyl 2,5-furandicarboxylate has been gaining notoriety within the last couple of years, mostly because conventional methods adapted from the synthesis of other polycondensates (such as polyesters) have been proved adequate for furan-based polyamides. Melt polymerization of dimethyl 2,5-furandicarboxylate with a mixture of cis and trans isomers of 1,3-cyclohexanedimethanamine were conducted in view of destroying the regular structure of final polymers and obtaining highly transparent materials [157]. Although spectroscopic analysis demonstrated the successful synthesis of polyamides, the reactions were prone to *N*-methylation and decarboxylation, which led to polymers of low molecular weights. A melt polycondensation method was also used to polymerize dimethyl 2,5-furandicarboxylate with varying diamines using low contents of titanium(IV) isopropoxide as catalyst, due to its excellent selectivity towards amidation reactions [158,159]. MALDI-ToF analysis indicated the presence of various end-groups apart from the expected amino and ester groups, which was attributed to side reactions that occur at the high temperatures employed.

To address this issue, prepolymerization followed by solid-state polymerization, a widely used strategy in the industrialization process of polyamides, have been performed using different catalysts. Poly(decamethylene terephthalamide/decamethylene 2,5-furandicarboxamide) copolyamides with furan ring contents ranging from 0 to 100 mol % were prepared from dimethyl 2,5-furandicarboxylate, terephthalic acid, and 1,10-decanediamine using this method, which was carried out in water as reaction medium and in the presence of sodium hypophosphite as catalyst [160]. Increasing contents of furan gradually impaired an amorphous character – and therefore an improved transparency – to the final materials, indicating that the intermolecular hydrogen bonds were suppressed. The relatively low values of Mn (5.2 to 7.7 kDa) and the wide molecular weight distribution (Đ values ranging between 2.40 and 3.80) were attributed to the solid-state polymerization step, carried out in a vacuum oven with no sufficient heat-transfer.

Recent contributions are related to the synthesis of poly(decamethylene furanamide) via bulk polymerization using 1,5,7-triazabicyclo[4.4.0]dec-5-ene (TBD) as catalyst. The authors obtained polyamides with Mn values up to 29.2 kDa but did not go further to investigate the occurrence of side reactions [161]. A further study of the same research group included a second step of solid-state polymerization aiming at enhancing the final polymer properties and avoiding/understanding relevant side reactions, but no scientific evidence was given besides possible theories involving the reaction of the continuously released methanol with reactive end groups of the polyamides [162].

Enzyme-promoted syntheses of polymers are of great interest due to its high selectivity and considerably reduced environmental impact. In this context, *Candida antarctica* lipase B (CALB) enzyme is often used for polyester and polyamide syntheses. As lipases can catalyze the formation of amide bonds, their use as catalysts for the synthesis of polyamides is attractive, but the incompatibility of such systems with the high Tm values and the low solubility of polyamides are yet to be addressed. At temperatures above the Tm (e.g., to perform melt polymerizations), the catalytic activity of lipases decreases significantly, which also happens in the presence of destructive solvents (concentrated sulfuric acid, formic acid, and trifluoroacetic acid), in which polyamides are mostly dissolved [59b]. Despite these issues, some interesting examples can be found in the literature dating from the first initiative related to the production of poly(octamethylene furanamide) catalyzed by Novozym 435 (an immobilized form of *Candida antarctica* lipase B) using a one-stage method in toluene (Mn values up to 13.4 kDa) and a temperature-varied two-stage method in diphenyl ether (Mn values up to 11.1 kDa) [163]. The same research group then expanded the strategy contemplating a wider range of aliphatic diamines differing in chain length (C4-C12), using the one-step method at 90 °C in toluene (Mn values up to 13.4 kDa) [164] and heteroatom diamines [165]. In all cases, the authors carefully elucidated the materials structures, with MALDI-ToF results indicating that the reactions mostly produced mixtures of polyamides with different ester/amine terminations, as well as macrocycles.

5.2.3 Polyamide from Furan Diamines

In a pioneering study which began in the 1980s [166], FDCA dichloride was polymerized in solution with several furan and aromatic diamines at the novel ensuing 26 polyamides fully characterized. The all-furanic polyamides were synthesized using a variety of diamines, including those shown in Figure 5.28 and structure/properties relationships established on the basis of the specific diamine structure used, through DSC and TGA data.

When aromatic diamines were employed [166], the most relevant result was associated with 1,4-phenylene diamine, which produced a Kevlar-type material where the aromatic acid moieties were replaced by 2,5-furandicarbonyl units (Figure 5.29)

Indeed, this polymer displayed all the typical features of a rigid polyaramid with a Tg of 325 °C, a thermal stability up to 445 °C, and solubility limited to oleum and NMP/LiCl. Another characteristic property was associated with the liquid-crystal behavior of its solution, with the classical critical concentration above which its viscosity deceases dramatically due to the alignment of the macromolecules and thus facilitates extrusion operations, i.e., fiber spinning. Polarizing light microscopy was used to construct a phase diagram of these solutions as a function of polymer concentration, separating the high-temperature isotropic phase from the liquid-crystal anisotropy at lower temperatures. This represents the first example of a furan polymer exhibiting all the features associated with polyaramids.

When other aromatic diamines were used as comonomers, the polymer rigidity was strongly reduced by the presence of flexible moieties separating two benzene

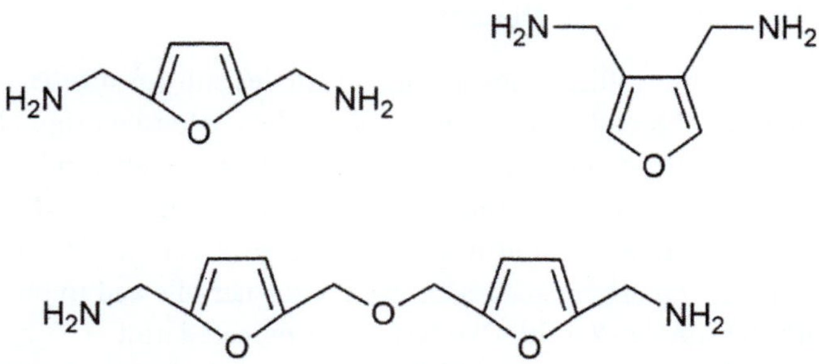

Figure 5.28 Furan-based diamines used for the synthesis of polyamides.

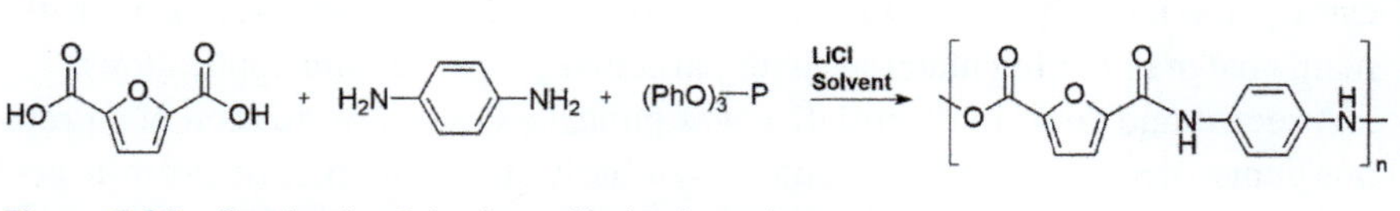

Figure 5.29 Synthesis of the furan-Kevlar homologue.

rings, but both the degradation and the glass transition temperatures remained high, albeit lower than those of the Kevlar-type homologue.

An interesting fact to be mentioned here is the paucity in novel contributions on furan-based diamines along the three subsequent decades, which recently changed with the preparation of the peculiar monomers shown in Figure 5.30 [167–170].

Due to the sophisticated molecular structures of the diamines depicted above, it becomes clear that the situation has found a favorable path that allows the design of increasingly specific furan-based materials with potential to be employed in precise applications. Poly(amide-imide)s are well known for their high thermal stability, being used in diverse fields of engineering, and when one considers their preparation from difuranic dihydrazides (Figure 5.29a) [167] and aromatic dianhydrides, gains in solubility, thermal and mechanical properties are expected. Optically active poly(amide-imide)s were prepared by the direct polycondensation of 3,5-diamino-N-(4-(5-(4,5-diphenyl-1H-imidazol-2-yl) furan-2-yl)phenyl)benzamide with two optically active dicarboxylic acids (Figure 5.29b) [169]. The materials were soluble in polar organic solvents at room temperature, which allowed a simple processing via solution-cast from DMAc solution into transparent, flexible, and tough films, capable of removing Cu^{2+} from aqueous solutions at pH 5. The AB_2-type monomer depicted in Figure 5.30c. was synthesized from furfurylamine and levulinic acid for the production of hyperbranched polyamides via melt polycondensation [168]. A further contribution from the same authors suggested future applications as ion-exchange materials [170].

5.2.4 Polyamides from Other Furan Monomers

Towards the end of the previous millennium, furan polyamides still constituted a minor subject, but the plan of broadening the possibilities by exploiting original monomers was already in vogue. A very clever idea was based on the self-condensation of 2-furamide (Figure 5.31) [171], similar to what was largely essayed to polymerize furfuryl alcohol: while the latter reaction produces –CH_2–2,5-Fu– oligomers that start a series of important side reactions, the former allowed to produce a polyamide with the simple –2,5-Fu–CO–NH–CH_2– repeated unit.

Much more recently, after some decades of hibernation, original furan-based monomers have now been presented either to demonstrate simpler and more efficient paths for polyamides that may be considered for wider uses, or with the main goal of obtaining materials with particular properties and applications.

A recent and very smart initiative was published as a continuation of a previous demonstration of a viable route of synthesis of 5-aminomethyl-2-furoic acid from CO_2 and furfuryl amine [172], which is its direct polymerization to

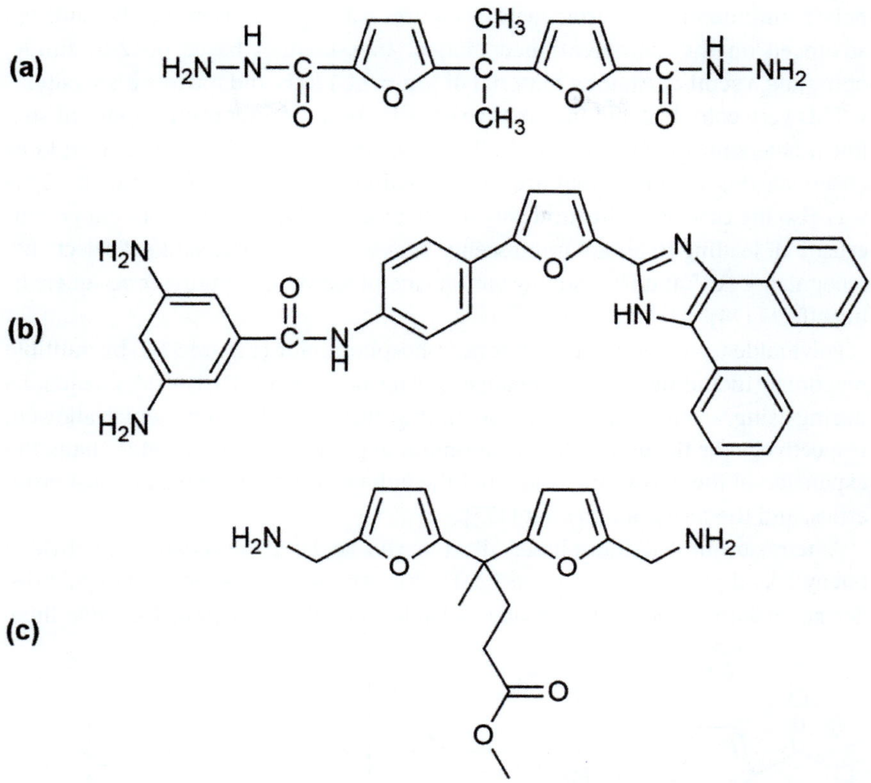

Figure 5.30 Molecular structures of some original furan-based diamines used for the synthesis of polyamides: (a) difuranic dihydrazide, (b) 3,5-diamino-N-(4-(5-(4,5-dephenyl-1H-imidazol-2-yl)furan-2-yl)phenyl) benzamine, and (c) 4,4-bis[5-(aminomethyl) furan-2-yl]pentanoate.

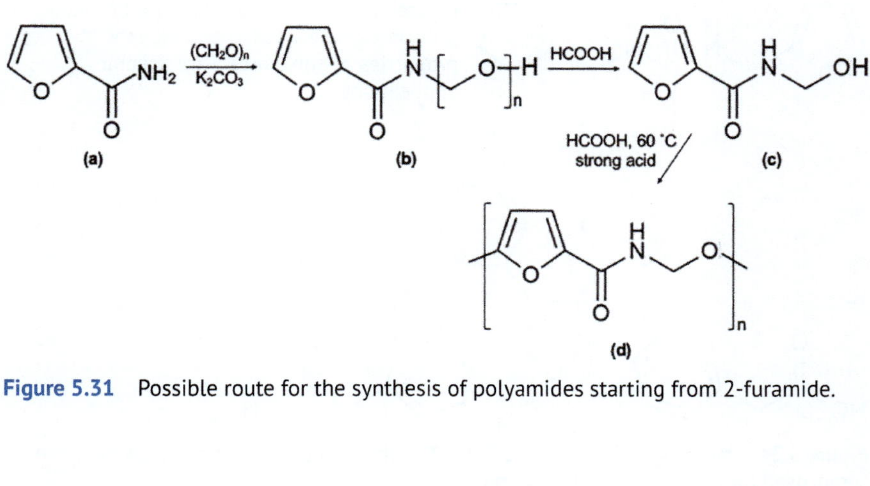

Figure 5.31 Possible route for the synthesis of polyamides starting from 2-furamide.

poly(5-aminomethyl-2-furoic acid) (Figure 5.32) [173]. Herein, the authors advanced on the aforementioned pioneer investigation based on 2-furamide, obtaining a semi-crystalline material of Mn = 11.1 kDa and narrow dispersity.

This very complete and informative contribution represents an important step towards technically and economically viable furan-polyamides, since it explores cheap starting materials that are easily obtained from biomass feedstocks. This was also the case when furfural was subjected to the Ugi four-component polymerization, leading to photoluminescence active polyamides capable of selectively recognizing Fe^{2+} and Fe^{3+} among various metal ions by the fluorescence quenching effect [174].

Polyamide 6 was modified with furan-phosphoramide (Figure 5.33) by multiple reactions (including amide exchange during processing, Diels–Alder reactions during using, aromatization and crosslinking during combustion), which allowed, respectively, the fixation of the flame-retardant group on the polyamide chain, the expansion of the polyamide chain and the enhancement of the mechanical properties, and the formation of char [175].

A tetra-substituted furan-based dicarboxylic acid (2,5-bis(4-carboxymethylene phenyl) 3,4-diphenyl furan, Figure 5.34) was synthesized and submitted to polycondensation with various commercially available aromatic diamines (4,4'-oxydianiline,

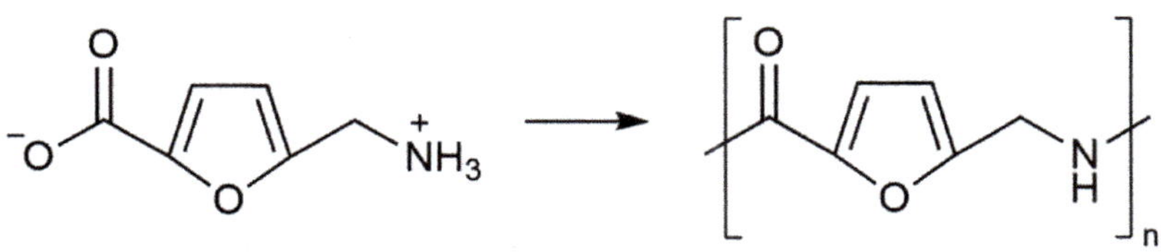

Figure 5.32 Synthesis of poly(5-aminomethyl-2-furoic acid) via organic phosphite and solid state polymerization.

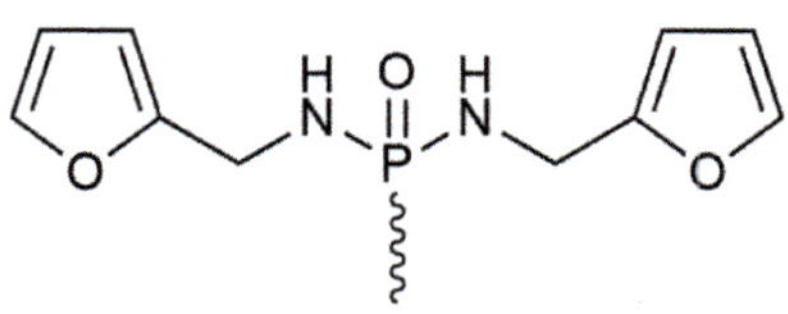

Figure 5.33 Molecular structure of furan-phosphoramide, used to enhance the properties of polyamide 6 for specific applications.

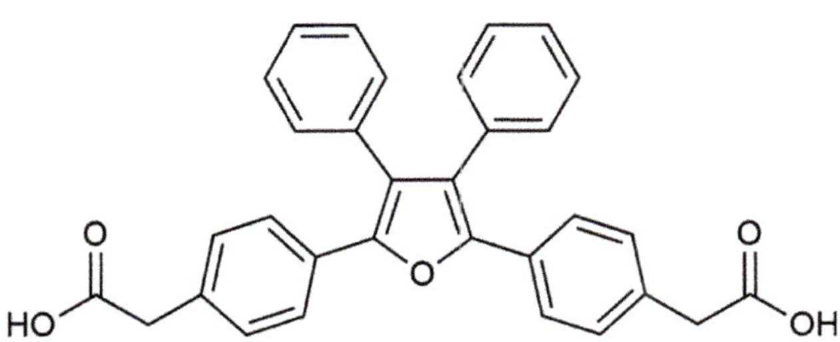

Figure 5.34 Molecular structure of 2, 5-bis(4-carboxymethylene phenyl) 3,4-diphenyl furan, used for the synthesis of polyamides.

4,4'-methylenedianiline, 4,4'-sulphonyldianiline, 4,4'-diaminobenzophenone and 4,4'-(hexafluoroisopropylidene)dianiline) in the presence of LiCl in NMP, pyridine, and TPP, at 100 °C for 3 h [176]. The authors reported that the series of four polyamides were thermally stable and organo-soluble, and suggested advanced applications, such as constituents for superior high thermal resistance devices.

5.2.5 Thermally Reversible Polyamides via Furan-maleimide Diels–Alder Reaction

The furan-maleimide Diels–Alder reaction applied within the context of furan polymers are thoroughly discussed in Section 6, with particular emphasis to the versatility it brings for tunnelling the properties of polymeric materials. As furan polyamides are advancing fast in both scientific and technological spheres, it is not surprising to find successful examples of thermally reversible polyamides via the furan-maleimide Diels–Alder reaction in literature. Considering the general topic will be tackled in detail in a further session of this book, only a couple of examples will be cited here.

Polyamides bearing various maleimide and furan pendant groups, as well as their crosslinked counterparts, were synthesized as represented in Figure 5.35 [177]. As expected, the linear polyamides behaved as thermoplastic materials and, after crosslinking, properties such as re-fabrication, high toughness, and self-repairing were attained. The synthesis of a series of comb-shaped polyamides with pendant furan groups (Figure 5.36), as well as the subsequent conversion into recyclable thermosets by DA reaction, were reported as a facile way of tailoring the mechanical properties of polyamide thermosets within broad range by adjusting the ratio of furan to bismaleimide moieties within the network [178]. A furfurylamine-N,N'-bis(methyl propionate) (Figure 5.37) was

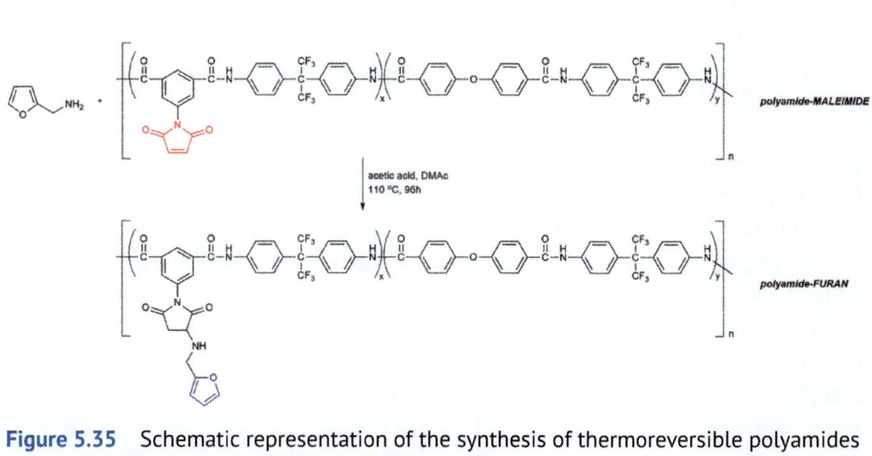

Figure 5.35 Schematic representation of the synthesis of thermoreversible polyamides based on the furan-maleimide Diels–Alder reaction.

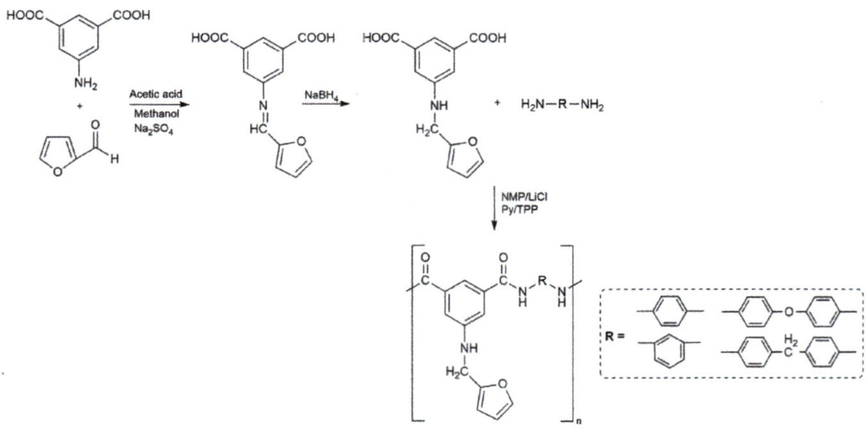

Figure 5.36 Synthesis and polymerization of a furan-containing diacid monomer used for the synthesis of thermoreversible polyamides.

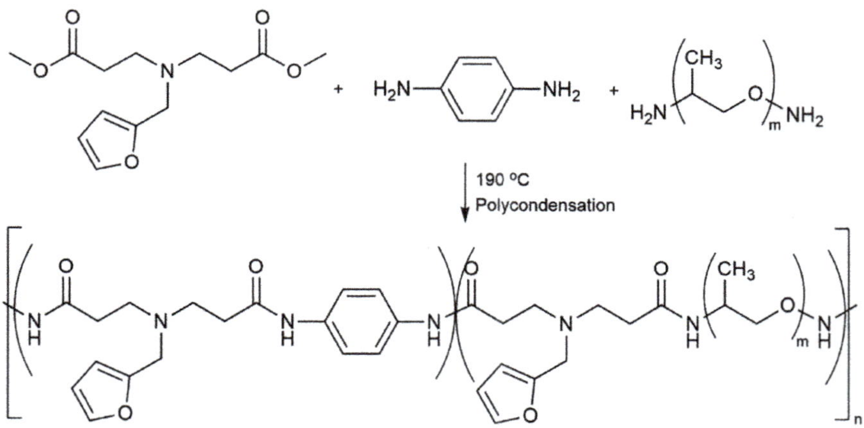

Figure 5.37 Schematic representation of the synthesis of furan-pendant polyamides from furfurylamine-N,N'-bis(methyl propionate).

synthesized through the Michael addition of furfurylamine and methyl acrylate. Polyamide prepolymers containing furan pendant groups were synthesized by bulk polycondensation of the diester monomers with bis(2-aminopropyl) polypropylene glycol and *m*-xylylenediamine. A further DA reaction was carried out using 1,5-bismaleimido-2-methylpentane, and the thermally repaired materials showed tensile strength up to 26 MPa and repairing efficiency of 70–92% [179,180]. A maleimide-designed polyhedral oligomeric silsesquioxane was associated with an aromatic polyamide bearing furan pendant-groups, and the final composite materials presented a good potential to be applied in electronic devices [181].

5.3 Polyurethanes

The domain of furan-based polyurethanes has interested polymer chemists and practitioners since the 1980s and is witnessing today a lively activity after a long lull. Before tackling the actual systems leading to these materials, it is useful to deal first with the relevant monomers, both isocyanates and polyols, and indeed with the mechanisms and kinetics related to the formation of furan monourethane model compounds. 2-furyl isocyanate (Figure 5.38), prepared from 2-furoic acid by the Curtius reaction, is extremely reactive in bulk, even when kept under an inert atmosphere at low temperatures, turning into a resinous material. It is, however, stable for weeks when stored in chlorinated hydrocarbons, as shown by the persistence of a clean ^{1}H NMR spectrum [182]. 2-furfuryl isocyanate (Figure 5.39), prepared from the reaction of 2-furfuryl amine with triphosgene, is much more stable, but is best kept, after purification, in chloroalkene or chloroarene solutions below room temperature under a nitrogen atmosphere [182].

After a thorough characterization of these two furan monoisocyanates, a systematic comparative study was conducted on the kinetics of the formation of a series of monourethanes (Figure 5.40), which included the use of isocyanate and

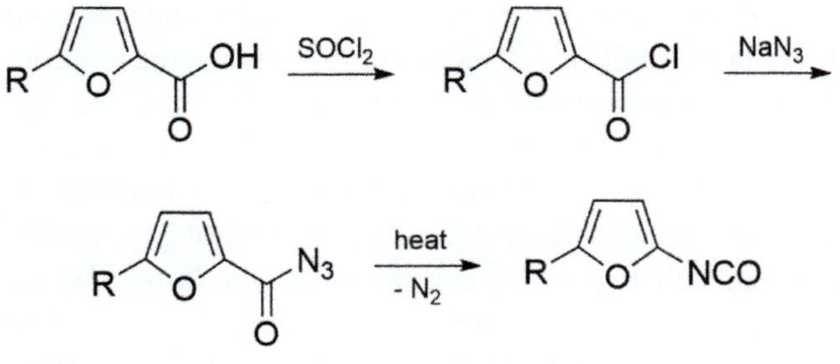

Figure 5.38 The synthesis of 2-furyl isocyanates.

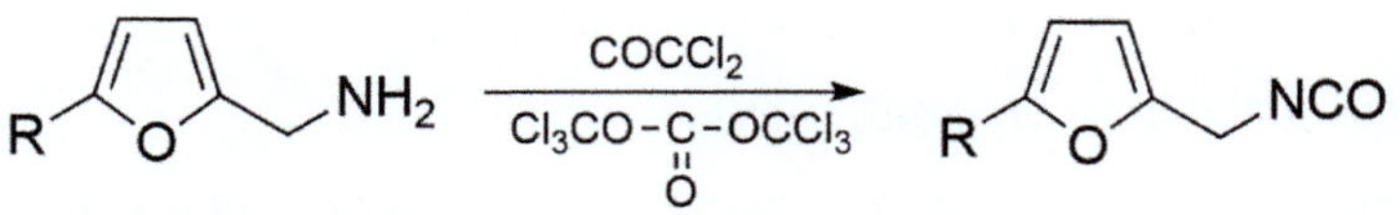

Figure 5.39 The synthesis of 2-furfuryl isocyanates.

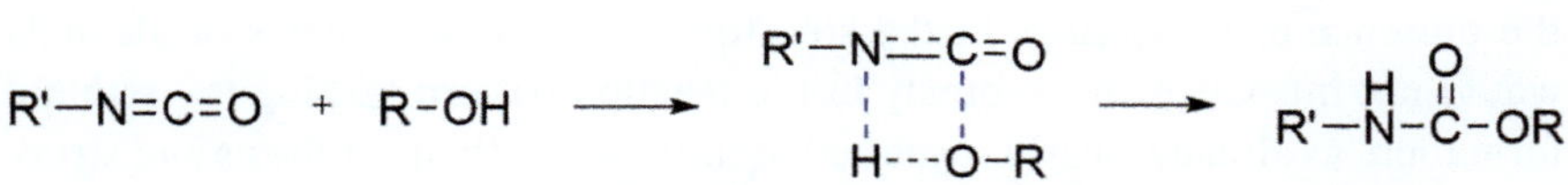

Figure 5.40 The mechanism of formation of monourethanes where – NCO is the electrophile and – OH the nucleophile.

alcohol reagents both incorporating furan moieties and aliphatic and aromatic counterparts, since this approach had not been investigated before [182,183].

Reactions were carried out in 1,2-dichlorobenzene at 110 °C under nitrogen with $[NCO]_0=[OH]_0= 0.5$ M and followed by GLC and all their kinetic patterns were found to follow a second-order behavior, viz.

$$-\frac{d[OH]}{dt}=-\frac{d[NCO]}{dt}$$
$$= k_2[OH][NCO]= k_2[OH^{-2}$$

from which the corresponding rate constant allowed useful reactivity comparisons on the basis of steric and electronic considerations. Thus, the reactions of phenyl isocyanate and *n*-butyl isocyanate with a series of alcohols gave the results shown in Table 5.5, which highlight two main features, viz. the general slowing effect steric crowding and the higher reactivity of the aromatic isocyanate. Additionally, the neighboring effect of both the phenyl and furyl moieties reduced the reaction rates by their electronic role, more so with the latter diene. On the other hand, the 5-methyl and 5-aldehyde substitutions on FA produced an increase in reactivity thanks to their donor character. Finally, furoin displayed the lowest reactivity given the enclosed position of its secondary -OH group. The reactivity of 2-furyl isocyanate towards all tested alcohols was extremely pronounced, to the point that no proper kinetic study could be conducted. This striking increase in the rate constants, estimated at two orders of magnitudes compared with that of phenyl isocyanate, was attributed to the dienic character of the furan ring providing a strong conjugation with the -NCO group, as indeed already reported for vinyl isocyanate. Turning now to the comparative activity of 2-furfuryl and benzyl isocyanates, Table 5.6 shows the systematically higher rate constants with the former, which suggests that the activating role of the heterocycle is operating through a field effect even when it is separated from the -NCO group by a methylene moiety. From these data, the following reactivity sequences could be established, namely:

$$nBuOH(pK_a = 16.1) > sBuOH(pK_a = 17.6) > tBuOH(pK_a = 19)$$

$$FuCH_2OH > FuCHMeOH > FuCMe_2OH$$

$$FuNCO \gg FuCH_2NCO > PhNCO > PhCH_2NCO > nBuNCO > tBuNCO$$

This kinetic assessment [183] was deemed important in order to confirm that the presence of furan rings in the structure of either isocyanates or alcohols would not introduce any anomaly in the reaction pattern leading to urethane formation, as already clearly suggested in the study of their synthesis and structural characterization [182]. In fact, the only quantitative aspect that was revealed here was the considerably higher reactivity displayed by the furan isocyanates even with respect to the well-known pronounced reactivity of their

Table 5.5 Second-order rate constants and domain of validity of the second-order behavior for reactions involving two isocyanates and a series of aliphatic aliphatic/aromatic and aliphatic/furanic alcohols. FE 1-(2-furyl)-1-ethanol; FP 2-(2-furyl)-2-propanol; MFA 5-methylfurfuryl alcohol. Fu stands for 2-substituted furan.

	Isocyanate			
	Ph-N=C=O		nBu-N=C=O	
Alcohol	**10^4 k_2 (1/mol s)**	**Domain of validity (% yield)**	**10^4 k_2 (1/mol s)**	**Domain of validity (% yield)**
*n*BuOH	37.6	90	5.0	70
*s*BuOH	10.5	75	1.4	70
*t*BuOH	3.6	75	0.6	75
*i*BuOH	26.0	90	-	-
*t*Bu-CH_2OH	20.5	95	-	-
*t*Bu-CH_2CH_2OH	36.7	90	-	-
Ph-CH_2OH	8.3	75	2.0	70
Ph-CH(Me)OH	1.8	70	0.5	45[a]
Ph-$C(Me)_2OH$	0.4	40[a]	-	-
Ph_2CHOH	0.04	15[a]	-	-
Ph_3COH	No reaction after 24 h	No reaction after 24 h	-	-
Fu-CH_2OH	5.0	75	0.7	75
Fu-CH_2CH_2OH	29.1	90	-	-
FE	3.2	80	0.5	75
FP	0.5	75	0.07	25[a]
MFA	7.1	70	1.3	75
HMF[b]	9.5	85	3.3	75
Furoin	0.5	75	0.07	25[a]

a) Reaction stopped at low conversion because of its slowness but still following a second-order law.
b) Hydroxymethylfurfural.

aromatic counterparts, a feature that was quite relevant for the forthcoming and fundamental investigation related to the preparation of furan-based polyurethanes discussed below.

The third phase of this comprehensive strategy was devoted to the synthesis and reactivity of furan diols and diisocyanates for the preparation of the corresponding diurethanes as models for the programmed polymerizations [184]. Figure 5.41 shows these compounds, which were synthesized or obtained as described in that publication and whose spectroscopic characterization confirmed all expected structures [184].

Table 5.6 Second-order rate constants and domain of validity of the second-order behavior for reactions involving 2-furfuryl and benzyl isocyanates and a series of aliphatic aliphatic/aromatic and aliphatic/furanic alcohols. MFA 5-methylfurfuryl alcohol.

	Isocyanate			
	Fu-CH_2-N=C=O		Ph-CH_2-N=C=O	
Alcohol	**10^4 k_2 (1/mol s)**	**Domain of validity (% yield)**	**10^4 k_2 (1/mol s)**	**Domain of validity (% yield)**
nBuOH	75.2	90	15.3	75
sBuOH	23.9	80	5.8	70
Ph-CH_2OH	21.4	85	3.8	80
Ph-CH(Me)OH	-	-	0.9	75
Fu-CH_2OH	9.4	90	2.6	70
MFA	15.2	80	3.7	70
HMF[a]	18	95	4.1	75

a) Hydroxymethylfurfural.

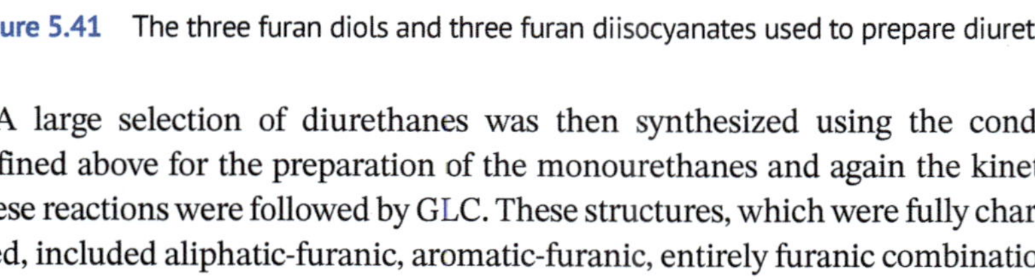

Figure 5.41 The three furan diols and three furan diisocyanates used to prepare diurethanes.

A large selection of diurethanes was then synthesized using the conditions defined above for the preparation of the monourethanes and again the kinetics of these reactions were followed by GLC. These structures, which were fully characterized, included aliphatic-furanic, aromatic-furanic, entirely furanic combinations, as well as conventional counterparts based on aliphatic and aromatic reagents for the sake of comparing behaviors, using the corresponding different monoisocyanates as co-reactants [184]. The kinetics of these numerous systems pointed to the same steric and electronic effects already found in the monourethane study [183], which led to the clear-cut general conclusion that no anomalous behavior could be detected concerning the presence of the heterocycle in this wide-ranging model survey and hence to proceed confidently to the work on furan polyurethanes. Before discussing this final phase, it is important to emphasize that other groups had investigated the synthesis of furan polyurethanes, but without the basic introductory approach just described here and or the wide range of structures discussed below. Apart from a rather qualitative report on the preparation of aliphatic and aromatic polyurethane foams and elastomers using BHMF and its oligomers as furan diols [185], only one other group researched furan-containing polyurethanes in the 1980s, including the synthesis of furan diisocyanates, and prepared various polymers based on difuran diisocyanates like those in Figure 5.41 and aliphatic diols [186]. Their thorough characterization led the authors to the conclusion that the replacement of conventional aromatic diisocyanates by the furan counterparts did not alter appreciably the thermal and mechanical properties of the ensuing materials. They also synthesized the first thermoplastic elastomer based on furan urethane segments and poly(ethylene oxide)diols potentially suitable for RIM technology [186].

The pursuit of the more comprehensive series of investigations outlined above finally reached the essence of the strategy by tackling the synthesis and characterization of a large domain of furan-based polyurethanes [187]. The furan monomers used in this wide-ranging survey included those shown in Figure 5.41 complemented by 3,4-bis(hydroxymethyl)furan, bis[5(hydroxymethyl)-2-furfuryl]ether and 2,5-furfuryl diisocyanate, as well as with some conventional aliphatic and aromatic comonomers. DMA solution polymerizations were carried out in a nitrogen atmosphere at 70 °C and some were followed kinetically through the time decrease of the intensity of the FTIR -N=C=O peak at 2250 cm^{-1}. The isolated materials were fully characterized in terms of structure, molecular weight, and thermal and mechanical properties [187]. Most of these polymers crystallized and could be annealed to their glassy state. Figure 5.42 shows the

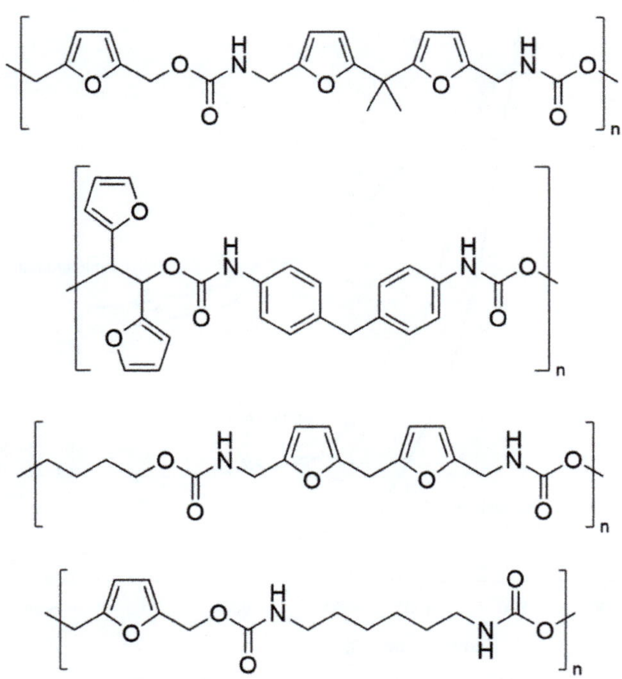

Figure 5.42 The structures of a selection of furan polyurethanes.

structures of a choice among the 15 furan polyurethanes prepared in this work, while their molecular weights varied between 5,000 and 20,000 Da. One of the relevant outcomes of this study is the fact that the thermal transitions of the materials could be readily modulated by (i) the position of the furan moiety in the chain (part of the backbone, side group and 2,5- vs. 3,4-substitution), (ii) the use of softening aliphatic units and (iii) the introduction of stiffening aromatic units. Thus, Tg values could be varied between 20 and 100 °C and Tm values between 150 and 210 °C. The TGA under nitrogen of these polymers confirmed an important feature related to the presence of furan motifs, already remarked in other macromolecular structures incorporating them, namely the extent of carbonaceous residue, as clearly illustrated in Figure 5.43, which reached some 45% with polyurethanes prepared with both furan diols and diisocyanates.

The final instalment of this series of papers moves into the realm of thermoplastic elastomers based on sequenced structures [188]. These materials are based on the conjunction low-Tg soft segments with high-Tg (or Tm) hard ones, which exhibit two distinct glass transition temperatures, typically well below and well above room temperature, so that their dynamic mechanical properties simulate

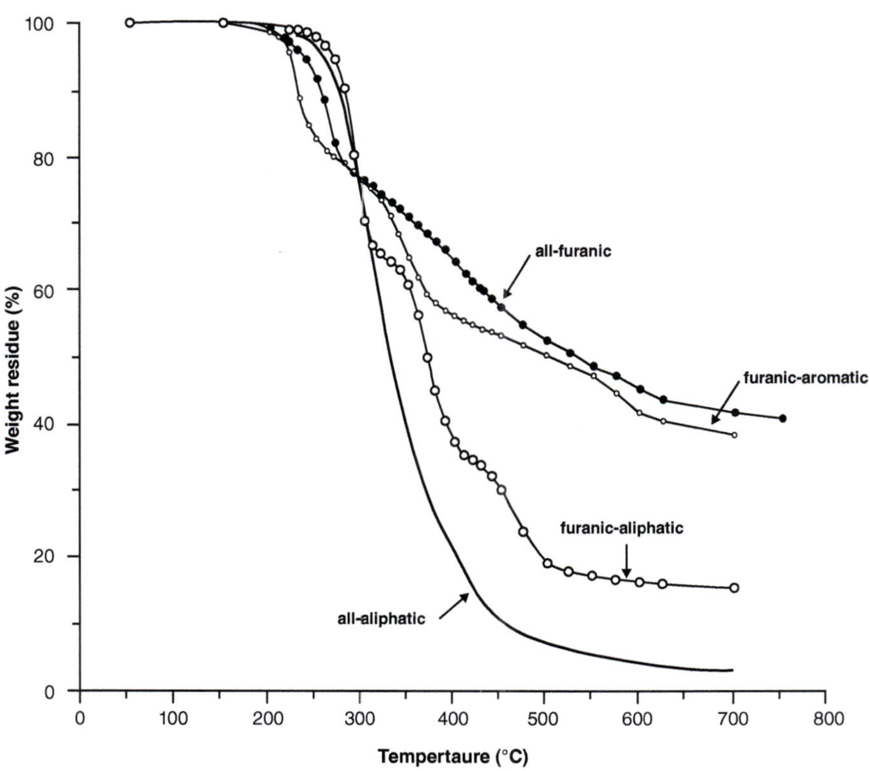

Figure 5.43 TGA tracings for different polyurethanes.

those of crosslinked elastomers. In this study, three low-Tg macrodiols were used for the soft segments, namely poly(tetramethylene oxide) glycol (PTMG), poly(caprolactone) glycol (PCL), and poly(butadiene) glycol (PB), with Mn of 2,000 to 3,000 Da. Their reaction with an excess of furan diisocyanates was followed by the addition of the furan diol chain extender in stoichiometric amount with respect to the remaining unreacted -NCO groups [188]. Both furan diisocyanates and diols were among those used in the previous investigation dealing with the polyurethanes discussed above [187]. Figure 5.44 illustrates the

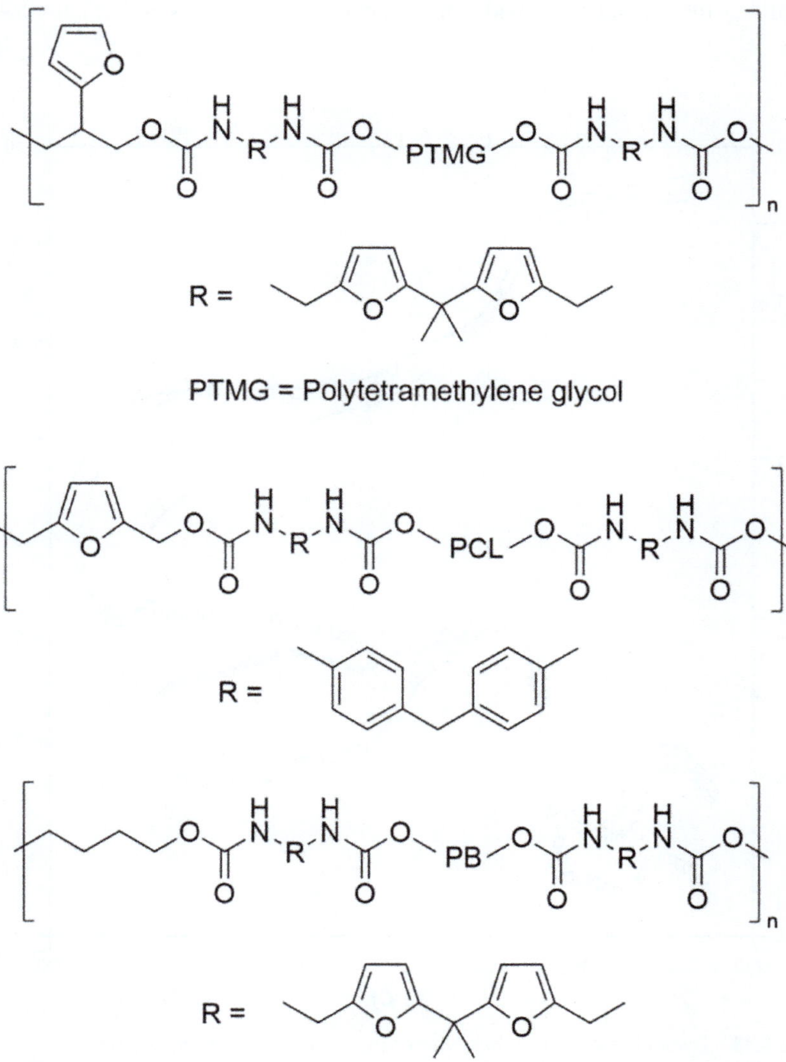

Figure 5.44 The structures of three typical thermoplastic elastomers using furan-based hard segments.

structures of some of the ensuing segmented elastomers. Their soft microdomain Tg values ranged between −20 and −70 °C depending on the specific reagents and the hard/soft segment ratios, whereas the Tg and/or Tm values referred to the hard microdomains went from 40 to 200 °C.

The dynamic mechanical properties of all these materials showed systematically a first drop in G', associated with the glass transition of the soft microdomains, followed by a plateau and finally a more dramatic drop in G', reflecting the polymer melting. Figure 5.45 shows examples of this behavior for two PTMG-based thermoplastic elastomers [188]. The stress-strain response of these materials could be modulated over a wide range of behavior, as illustrated in Figure 5.46,

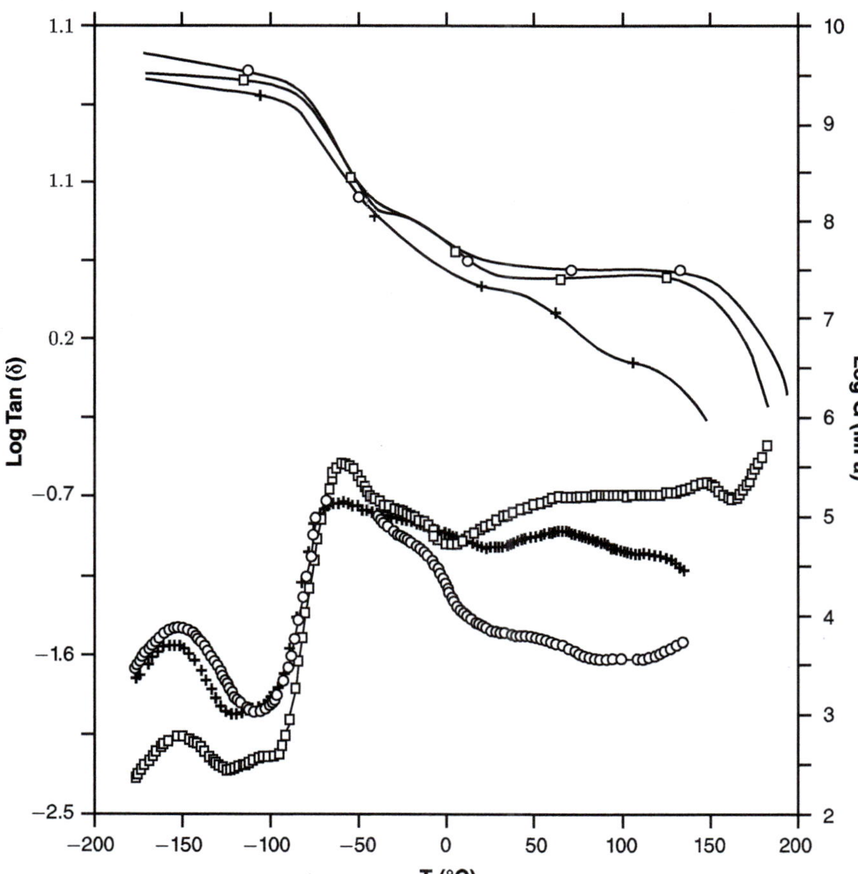

Figure 5.45 Dynamic mechanical behavior of three PTMG-based sequenced polyurethanes in which three different furan diols were used in conjunction with MDI.

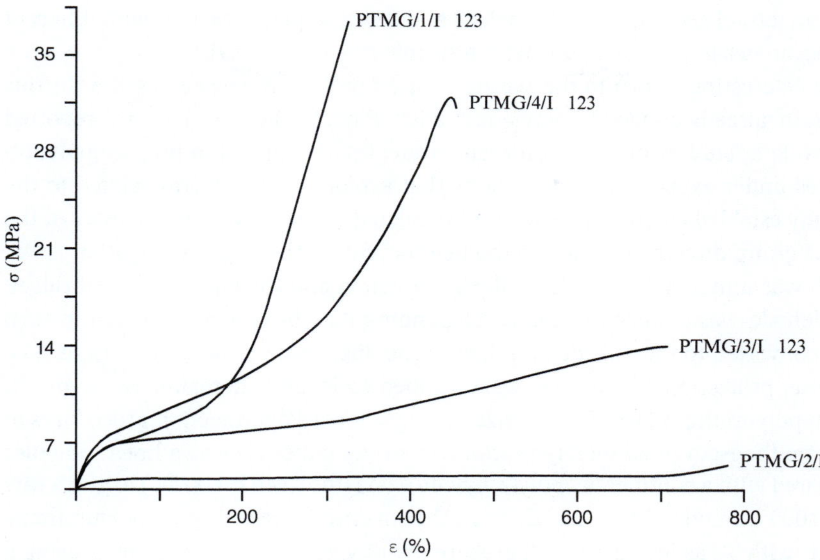

Figure 5.46 Stress-strain patterns for PTMG-based materials in which the diisocyanate was MDI (I) and the chain extenders BHMF (1), 1,4-butandiol (4), 3,4-bis(hydroxymethyl) furan (3) and 2-furyl-1-ethylene glycol.

where PTMG-based thermoplastic elastomers based on different chain extenders used at the same stoichiometry.

This investigation of thermoplastic elastomers based on furanic segmented polyurethanes provided a new family of materials which displayed moduli and thermal transitions covering domains of applications similar to those of aromatic-aliphatic counterparts. In more general terms, the deep incursion into the realm of furan-bearing polyurethanes [1–3,6,7] convincingly placed these novel materials as valuable siblings within the general domain of that macromolecular family, confirming the sustainable value of using monomers from renewable resources.

Rigid polyurethane foams were prepared from the furan macrodiol derived from 2-furyl oxirane shown in Figure 4.37 and a commercial MDI with functionality of 2.4, using standard additives to promote the formation of the alveolar morphology. The ensuing materials (Figure 5.47) displayed excellent features in terms of the percentage of closed cells, dimensional stability, and thermal conductivity, all comparable with those of commercial counterparts, including a much better stability against ageing [189]. Their additional original property was related to flame resistance, since intumescence generated by initial combustion rapidly extinguished the burning process, again mostly due to the high furan content, as discussed above (see Figure 5.43). Another approach called upon the synthesis of a furan tetraol according to Figure 5.48 and its subsequent reaction with different diisocyanates to prepare foams with tuneable properties [189].

An interesting return to the synthesis of 2-furyl isocyanates using the Curtius reaction, already applied in previous studies (Figure 5.38) [182,184] was reported in two later studies in which different strategies are applied to prepare polyurethanes under experimental conditions that avoided the problems related to the already established atmospheric fragility related to the very high reactivity of the -NCO group directly attached to the heterocycle. In the first investigation [190], HMF was dimerized through its alcohol function and the ensuing ether-bridged dialdehyde was oxidized to the corresponding dicarboxylic acid, which in turn was converted to the diazide and then to the diisocyanate shown in Figure 5.49. Various protection approaches were applied to it, and thereafter two types of furan polyurethanes were synthesized using standard deprotection procedures to liberate the isocyanate moiety *in situ*. One of the materials was a linear polymer prepared with a commercial poly(ethylene oxide) diol (Mn = 600 Da) to give a Mw of 50,000 Da and a Tg of −32 °C. The other materials were flexible polyurethane foams with Tg as low as −50 °C, prepared with various polyether polyols using a nitrogen flow to generate the alveolar architecture. This preliminary study was not followed, as far as we know, by a published systematic approach.

Figure 5.47 Comparison between the ageing effect on conventional rigid polyurethane foams (I) and the furan counterpart prepared from the diol in Figure 4.37.

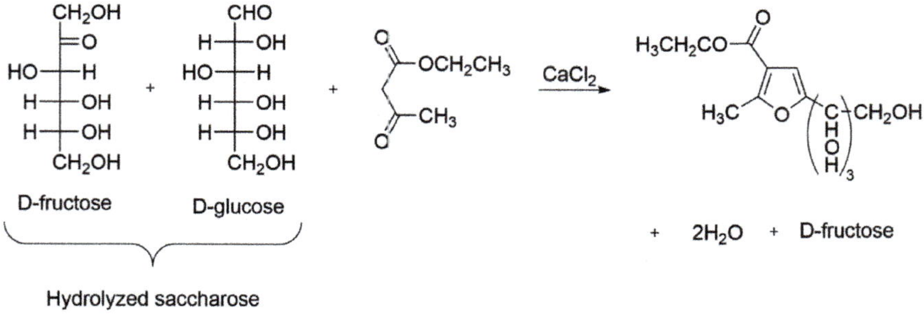

Figure 5.48 Synthesis of a furan polyol from saccharose.

The second work [191] used FDCA as the precursor to prepare the corresponding furandiacylazide (FDAz), which was used as a building block for two types of materials as shown in Figure 5.50. When used as such, it produced a network of poly(2,5-furanyl isocyanurate) arising from the cyclization of corresponding

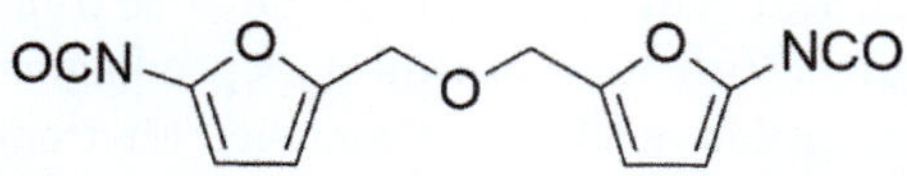

Figure 5.49 The furan diisocyanate derived from the HMF ether dimer.

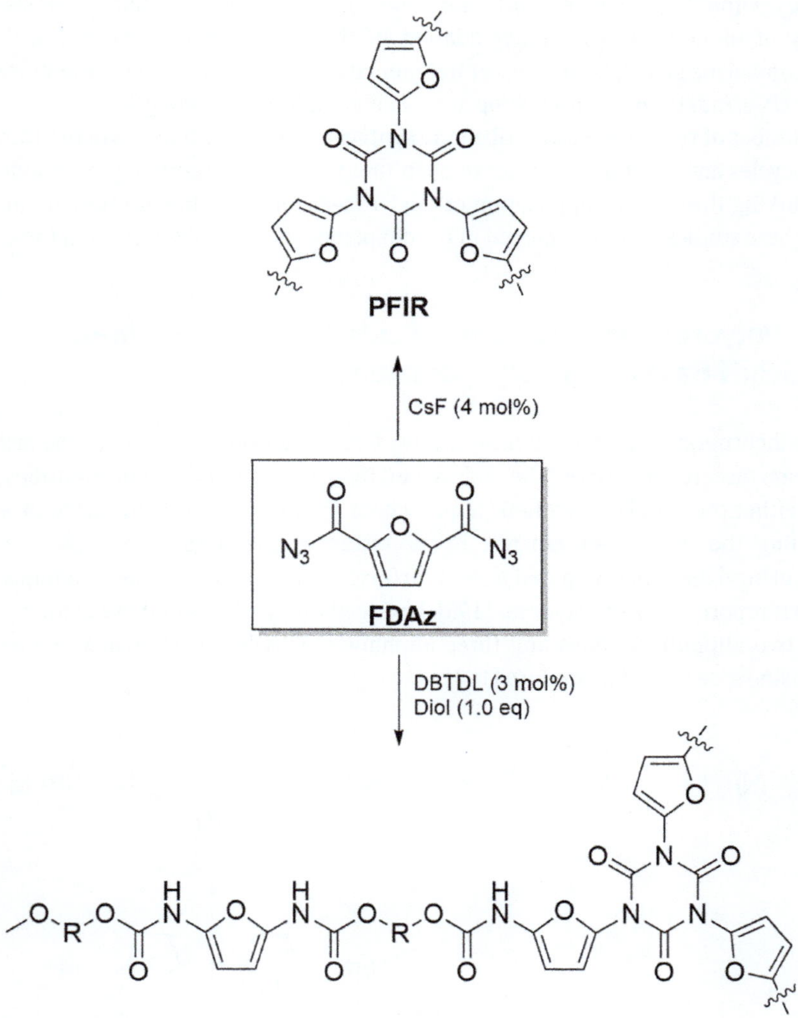

Figure 5.50 The synthesis of two furan polyurethanes from FDAz.

isocyanate, as already reported above in the study of the anionic polymerization of 2-furyl- and 2-furfuryl isocyanates (Figure 4.32) [192], with the major difference that here the difunctionality led to a crosslinked polymer (Figure 5.50, top structure) instead of simple trimers. When FDAz was used as comonomer with various diols, the ensuing materials had complex blocky structures made up of linear polyurethane sequences and branched motifs arising from the same trimerization process described above (Figure 5.50, bottom structure). Depending on the diol used, the Tg of these polymers ranged from 50 to 80 °C and their Mn from 20,000 to 50,000 Da. The most attractive aspects of this very thorough study were, on the one hand, the lack of flammability of the furan-rich crosslinked polytrimer (already pointed out on several occasions above), and, on the other hand, the possibility of inducing the photodegradation of the polyurethane prepared with 2-nitro-*p*-xylene glycol, i.e. a polymer fragmentation at the aromatic chromophore under UV irradiation, an interesting approach to material recycling.

A number of reports have described the synthesis of polyurethanes where furan heterocycles are appended in order to open the possibility of thermally reversible crosslinking through the application of the Diels–Alder reaction with bismaleimides. These studies will be discussed in the chapter devoted to that important topic.

5.4 Polyureas, Polyparabanic Acids, Polybenzoxazines, polySchiff Bases, and Polyhydrazides

Furan-incorporating polyureas have received little attention and only in the last few years researchers revived the topic, albeit in modest number of contributions, using either the standard synthetic approach, or the more sustainable alternative excluding the use of isocyanates. In all instances the furan monomers are 2,2-difurfuryl diamines prepared from 2-furfuryl amine and aldehydes or ketones. The first report on furan polyureas [193] described the synthesis of these diamines using two aliphatic ketones and three aliphatic aldehydes, as shown in Figure 5.51, using a described procedure [194].

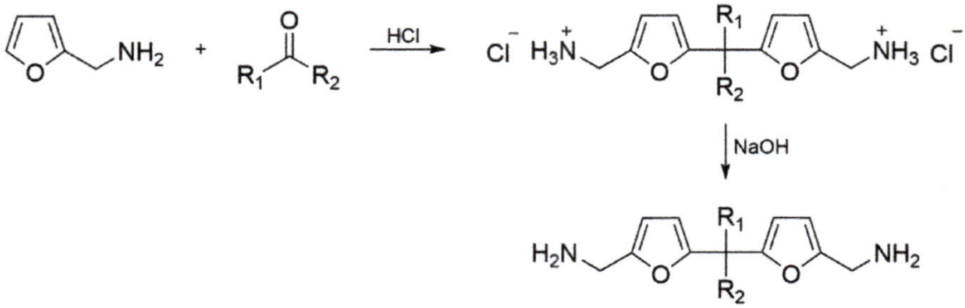

Figure 5.51 Synthesis of 2,2-difurfuryl amines with aldehydes and ketones.

Seven polyureas were then prepared by their standard polycondensations with three aliphatic diisocyanates, after an optimization of the conditions using 2-furfuryl amine as model for the synthesis of the corresponding diureas. The polymers were further modified through the cyclocondensation of the urea moiety by reacting them with oxalyl chloride, yielding the corresponding seven polyparabanic acids, as illustrated in Figure 5.52 for both types of macromolecular syntheses [193].

All these materials displayed the expected linear regular structures as verified by FTIR and ^{1}H and ^{13}C NMR spectroscopy and gave intrinsic viscosities of about 0.1 in *m*-cresol at 25 °C.

A more thorough study based on the same approach was later conducted in which the highly purified 2,2-difurfurylamine derived from acetone was reacted with hexamethylene diisocyanate and its 4,4-methylenediphenyl counterpart to prepare a furan-aliphatic (PMDI) and a furan-aromatic (PHDI) polyurea [195], as shown in Figure 5.53. The spectroscopic characterization of both polymers confirmed their structures with Tg values of 98 and 165 °C, respectively, reflecting the stiffening role of the aromatic units, and thermal stability in a nitrogen atmosphere exceeding 300 °C for both. Their static mechanical properties are shown in

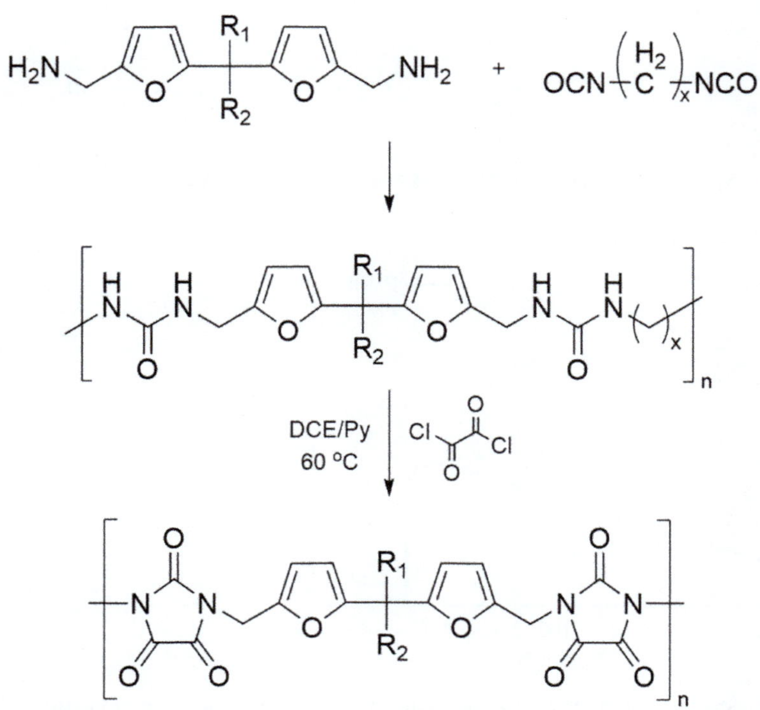

Figure 5.52 Synthesis of furan polyureas and polyparabanic acids.

Figure 5.54 as stress-strain plots, where the maximum tensile strength at break was 1.4 (± 0.4) MPa for PHDI and 2.3 (± 0.9) MPa for PMDI. The elongation at break was 148 (± 15) % for PHDI and 22 (± 9) % for PMDI. The elastic moduli were 4 (± 0.14) and 120 (± 8.70) MPa for PHDI and PMDI, respectively. These large differences are a clear reflection of the major role played by the aliphatic versus the aromatic moieties in determining the moduli and elongation responses of the polyureas in which they were incorporated, while the furan segments remained the same [195].

These polymers could be cast into films by hot pressing PMDI at 100 °C and PHDI at 165 °C for a few minutes.

As mentioned above, it is considered advisable to avoid the use of isocyanates in the preparation of polyurethanes and polyureas by either using protected

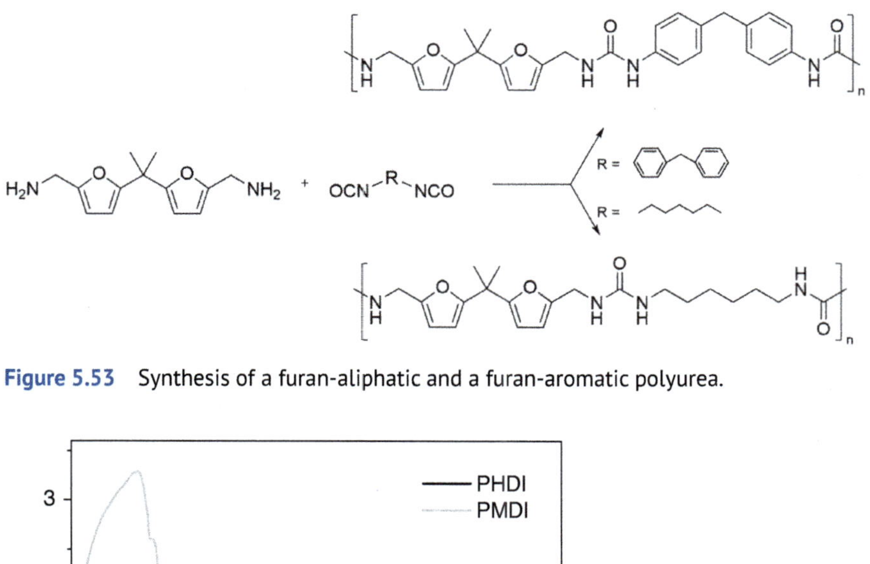

Figure 5.53 Synthesis of a furan-aliphatic and a furan-aromatic polyurea.

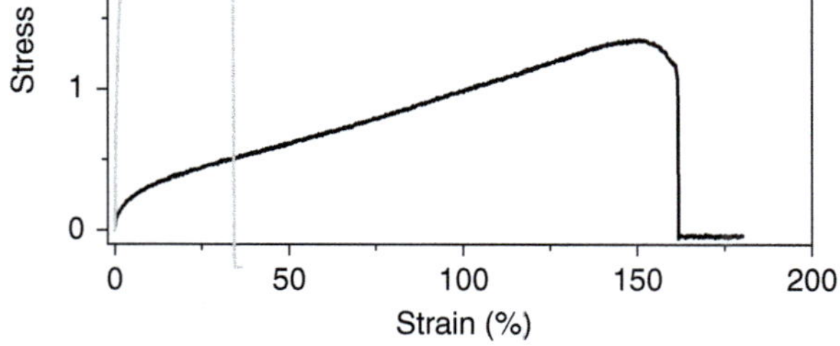

Figure 5.54 Stress-strain plots for the furan-aliphatic (PHDI) and the furan-aromatic (PMDI) polyureas.

forms or alternative synthetic routes often involving the intervention of CO_2 [196]. A very interesting example of the latter approach describes the preparation of a series of furan-based polyureas by the reaction of a furan dicarbamate with various diamines [197]. Here, the 2,2-difurfuryl amine in Figure 5.53 was reacted with dimethyl carbonate to prepare the corresponding dicarbamate, which was the furan monomer used to prepare different polyureas using a series of aliphatic and aromatic diamines following the scheme in Figure 5.55 [197]. Depending on the diamine used, these polymers had Mn values ranging from 3,000 to 17,000 Da and Tgs going from 11 to 164 °C, all being stable up to about 300 °C. It is regrettable that in all these studies on furan polyureas, none tackled the synthesis of a structure where both monomers bore the heterocycle, namely an all-furan polymer.

Polybenzoxazines (PBZ) have become over the years a strong alternative family of materials to their traditional counterparts like phenolic resins, phthalonitriles and cyanate esters thanks to their excellent properties such as high thermal stability, good mechanical performances, high glass-transition temperatures, rich structural tenability, near-zero volatile emission, and shrinkage during polymerization, high char yield, and very low moisture absorption [198]. Monooxazines, whose typical synthesis is shown in Figure 5.56 A, do not polymerize, but their dimer forms do so by ring-opening at high temperatures, as in the example of Figure 5.56 B. Given the possibility of innumerable potential substitution, a vast amount of work has been devoted to the elaboration of materials and their structure-properties relationships associated to the optimization of applications in different technological domains [198].

Recent advances have concentrated on developing monomers incorporating structures derived from renewable resources including lignin derivatives,

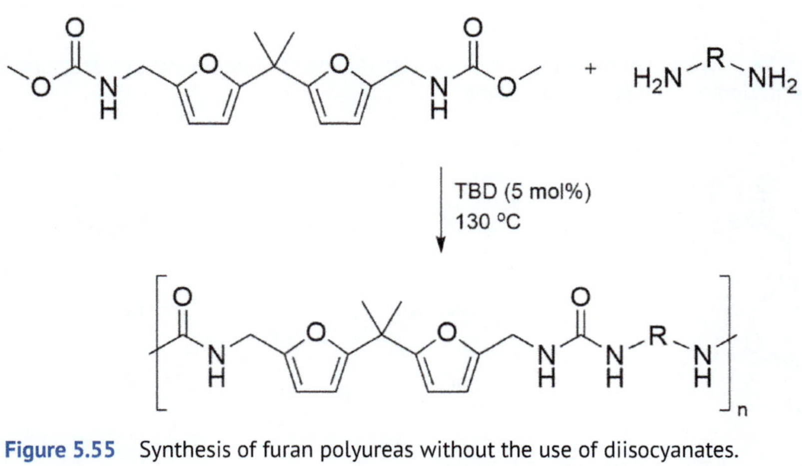

Figure 5.55 Synthesis of furan polyureas without the use of diisocyanates.

coumarin, cardanol, apigenin, and furan amines, among others [6,7]. The role of furan moieties in this context can be divided into two categories of materials, namely the use of 2-furfuryl amine as a coupling agent on the periphery of the oxazine [198–201], or the incorporation of 5,5-methylene-2,2-difurfurylmanine in the core of the monomer [198,202]. Figure 5.57 illustrates two typical monomer structures of the first type with respectively one and two furan heterocycles. The mechanism leading to crosslinking with the monosubstituted oxazine involves a cationic initiator, ring opening and electrophilic substitution, as shown in Figure 5.58. With two furan moieties, the final polymer will be a much denser network and hence with correspondingly different properties [198].

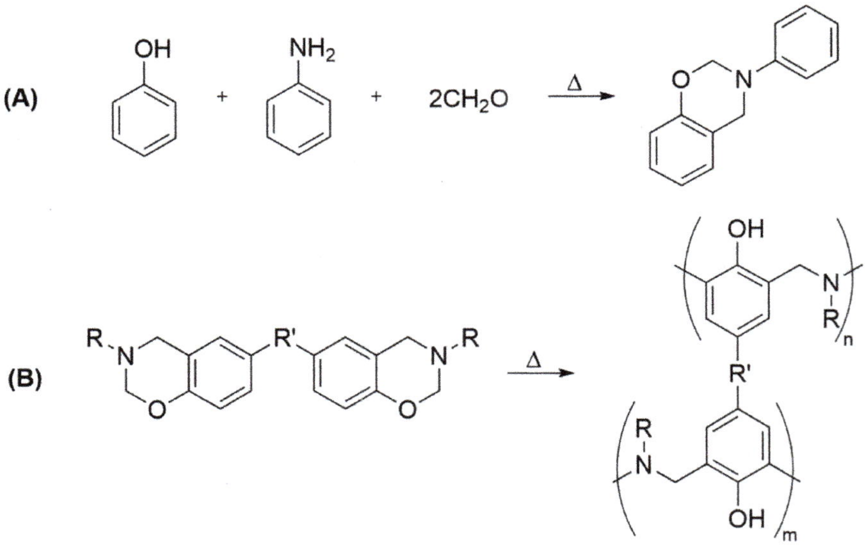

Figure 5.56 Typical synthesis of a mono-oxazine (A) and the ring-opening crosslinking polymerization of a generic oxazine dimer.

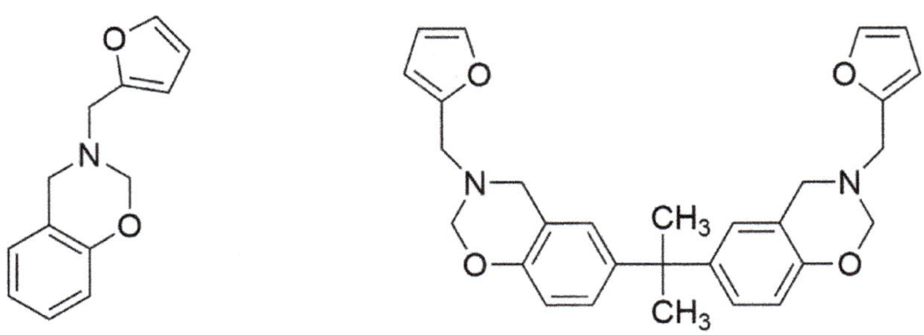

Figure 5.57 Typical furan-bearing oxazines.

A three-substituted oxazine has also been prepared and polymerized [201] following the proposed mechanism given in Figure 5.59. In all instances (Figures 5.58 and 5.59), as expected, the furan ring intervention is through the C5 substitution, confirming once again this vast regioselectivity with respect to the C3 and C4 positions.

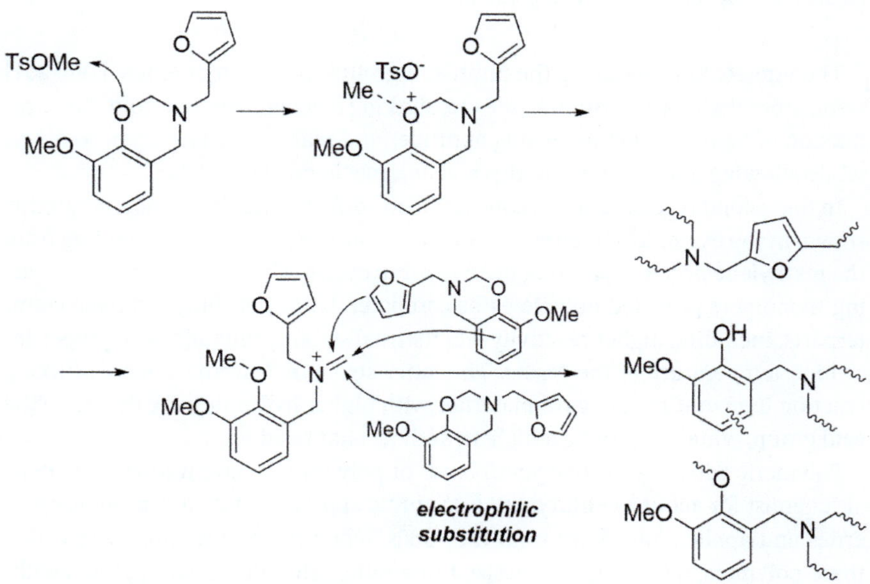

Figure 5.58 The possible polymerization mechanism of a furan oxazine monomer.

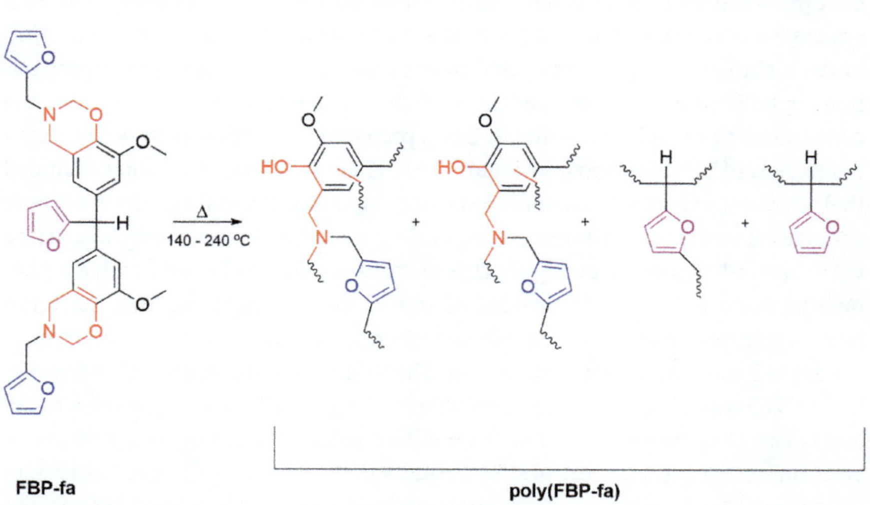

Figure 5.59 The types of crosslinked units present in the network arising from the polymerization of a trifuran oxazine.

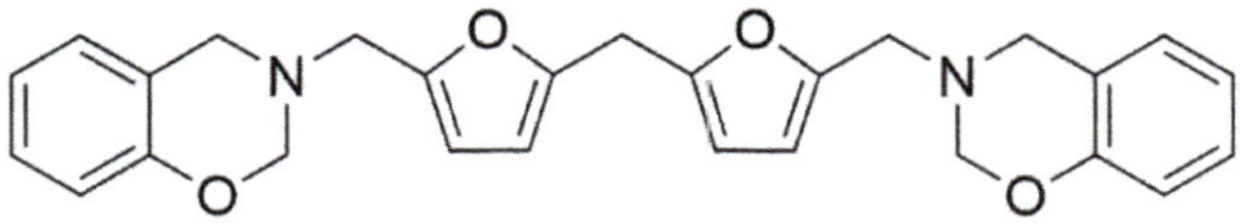

BZ-F

Figure 5.60 A benzoxazine incorporating a difuran moiety.

The interest in introducing the furan end-groups in these monomers [198–201] stems from their participation in the crosslinking mechanisms and from the introduction of flame self-extinguishing aptitudes and marine antifouling in coatings, while allowing a wider range of thermal and mechanical properties.

In the second type of furan oxazine [202], the difurfurylamine is incorporated as shown in the example of Figure 5.60, or with a carboxylic function extending from the methylene bridge separating the two furan cycles. The use of both furan-bearing monomers provided notable gains compared with their fully aromatic counterparts, including higher reactivity and thermal stability plus adhesive properties at high temperature of the resins. The latter structure lowered the crosslinking reaction temperature and gave materials with higher hydrophilicity thanks to the acid group, while preserving a high Tg and the char residue.

Polymeric Schiff bases (polySchiff bases or polyimines) have interested materials chemist for about a century, but a scientific approach to their synthesis, properties and applications dates from the 1960s, when it was also emphasized that these polymers, which are synthesized in a rather straightforward process without the need of a catalyst, possess high thermal stability, possible electronic conduction when fully conjugated and potential liquid-crystal behavior. The first systematic investigation dealing with the synthesis of furan polySchiff bases followed a similar strategy to that used in the progressive approach to the preparation of polyurethanes discussed in Section 5.3, namely by starting from the corresponding model mono- and di-bases prepared in solution with near-quantitative yields [203], as shown in Figure 5.61. Their full characterization included their electronic spectra that confirmed that compounds B and C (Figure 5.61) bore conjugated structures with the corresponding reduction in the energy associated with the π–π^* transition, giving absorption maxima close to 360 nm, with a higher melting point for C (170 °C instead of 160 °C for B), suggesting that the furan heterocycle introduced a higher cohesive energy than the aromatic counterpart.

Four polymers were then synthesized. The furan-aromatic structure E shown in Figure 5.62 was a brown powder insoluble in all tested solvents displaying a broad absorption maximum centered at 465 nm (KBr pellet), indicating a high degree of electronic conjugation, confirmed by a conductivity of about 10^{-4} S cm^{-1} after doping with iodine. Its FTIR spectrum suggested however a fairly modest DP because of the presence of a significant peak at 1670 cm^{-1} arising from carbonyl end groups.

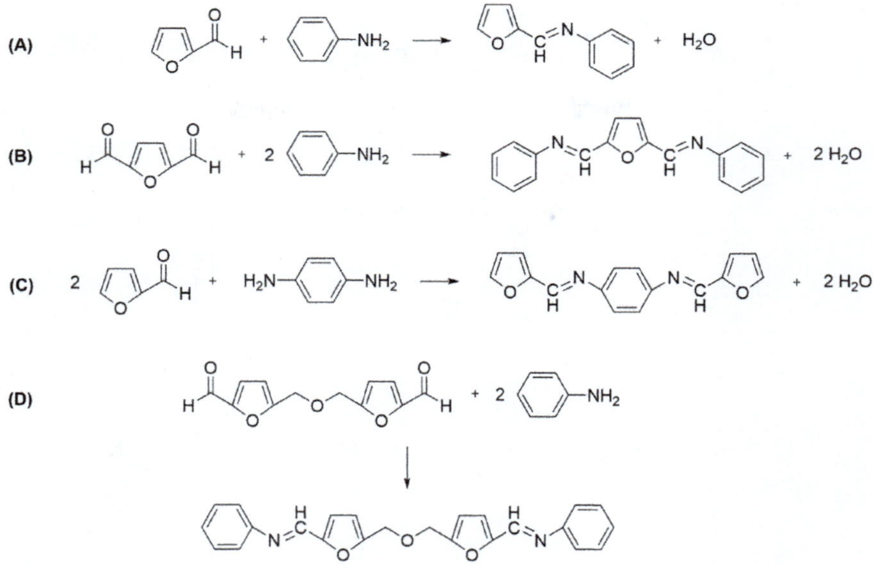

Figure 5.61 Model mono- and bis-Schiff bases from (A) aniline plus furfural, (B) aniline plus 2,5-furan dialdehyde, (C) 1,4-phenylene diamine plus furfural, and (D) aniline plus bis(2-formyl-5-furfuryl) ether.

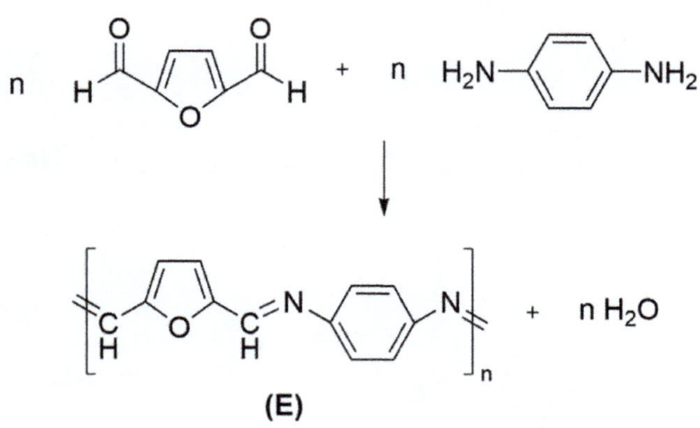

Figure 5.62 The fully conjugated furan-aromatic polySchiff base E.

Among the other three polymers, displayed in Figure 5.63, only F was white and soluble in some common solvents because of its long aliphatic units and hence an Mn of 2,500 Da could be measured, as well as a strong birefringence with a crystal melting at 148 °C. Polymers G and H were again insoluble in all tested solvents and displayed coloration (despite the lack of extended conjugation as in F) and

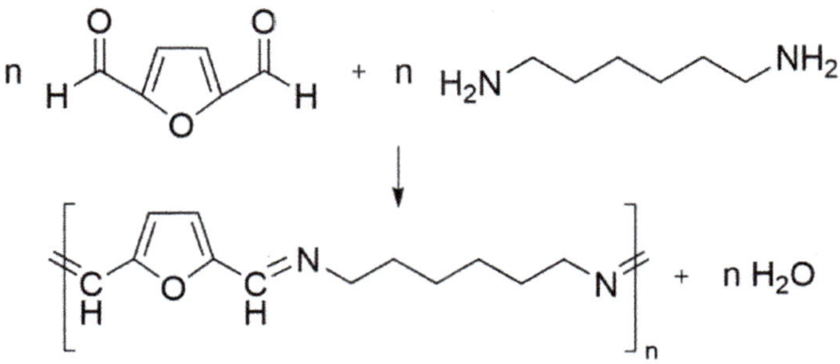

(F)

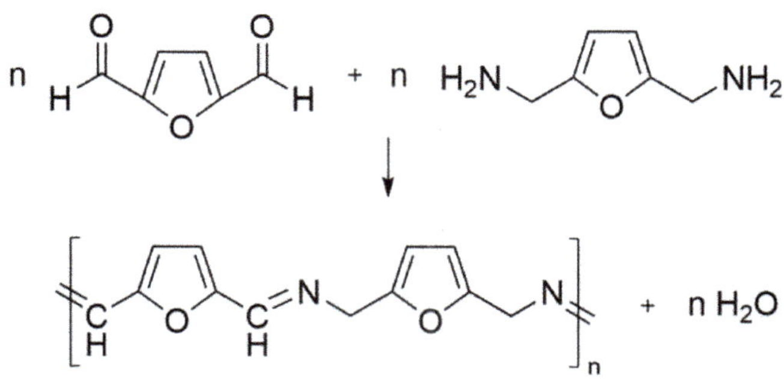

(G)

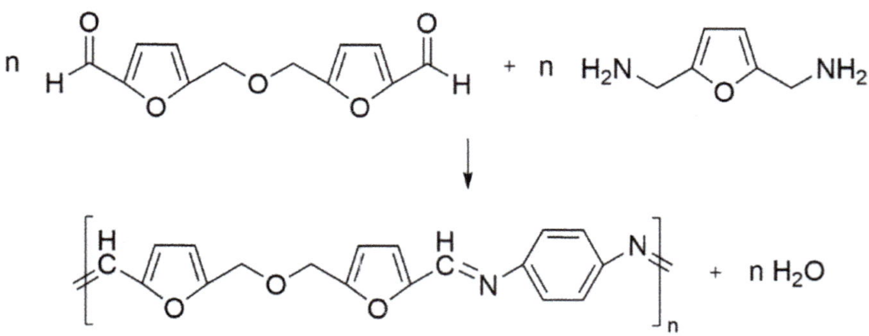

(H)

Figure 5.63 The other three furan-based polySchiff bases.

some thermal instability probably due to the methylene groups attached to the furan rings [203]. This structural feature was inevitable in the present context because furan amines with the -NH_2 groups directly attached to the ring are thermodynamically unstable and therefore cannot be envisaged as monomers.

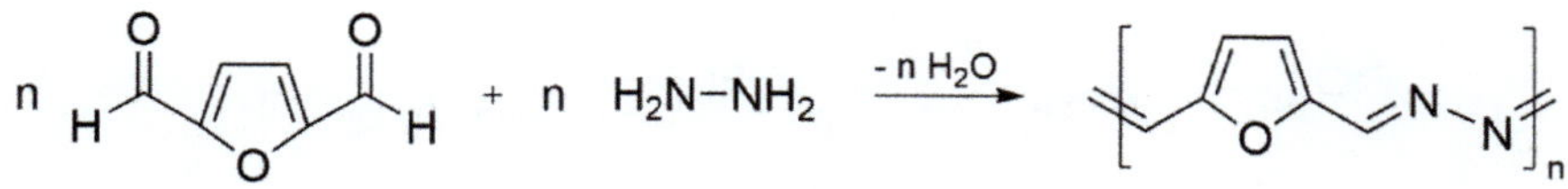

Figure 5.64 Synthesis of a furan polyazine.

In order to prepare another type of highly conjugated polySchiff bases, a second study was carried out based on the reaction of FDCA with hydrazine to give the corresponding polyazine, as sketched in Figure 5.64, after having prepared and fully characterized the corresponding model compounds. The polymer was sparingly soluble in methylene chloride, but sufficiently to verify spectroscopically that it bore the expected structure with an Mn of 2,000 Da. Its electronic spectrum gave a maximum at 440 nm and conductivities between 10^{-5} and 10^{-4} S cm^{-1} after doping with iodine. There was therefore no clear improvement on the properties of the materials, when switching from classical diamine [203] to hydrazine [204]. Interestingly, two subsequent studies on essentially the same system previously published [203] reached very similar results [205], except for pointing out that these polycondensations lead also to the formation of cyclic trimers.

The obvious drawback associated with developing viable furan polySchiff bases stems from their high cohesive energy which translates into very poor solubility and hence the formation of low molecular weights, because of premature precipitation during their synthesis. Two possible solutions were envisaged and indeed tested in preliminary experiments [204], namely introducing flexible blocks, e.g., oligoethers, within the polymer backbone, or appending short aliphatic strands onto the main chain, which however were not pursued to provide a confirmation of their validity. It seems therefore reasonable to consider further research on these topics, which appear worth of optimization. This was indeed pursued very recently, but since main purpose of this study was the application of the Diels–Alder (DA) reaction to the prepared soluble *oligomeric* Schiff base, the work will be dealt with in the appropriate DA section.

The significant step forward in the realm of conjugated furan polymers, compared with the first modest success related to polySchiff bases, will be discussed when dealing with the polycondensation of 5-methyl furfural in the next session.

Polyhydrazides are the final type of furan polycondensates involving diamines [206]. A series of furan dihydrazides were synthesized as shown in Figure 5.65, fully characterized, and then submitted to polycondensation reactions with diacid chlorides. The reactions were performed by both solution and interfacial procedures, as illustrated in Figure 5.66, after a preliminary study of the corresponding model compounds carried out in order to assess the optimized polymerization conditions.

All products were isolated in high yields and displayed clear spectroscopic evidence of their expected structures and the polymers gave inherent viscosities suggesting that the interfacial technique produced higher molecular weights [206].

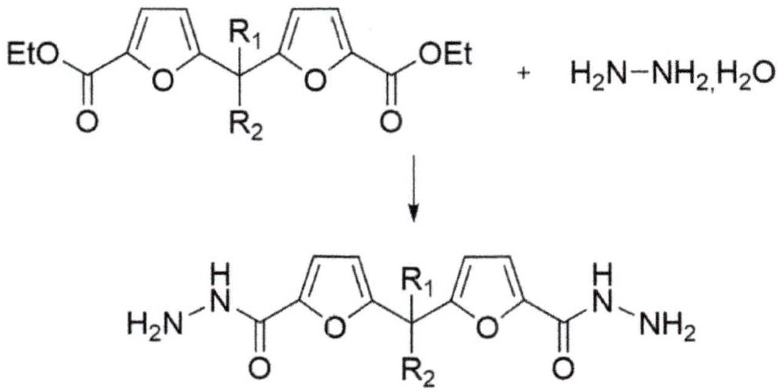

(1-9)

1: R_1,R_2 = CH_3, m.p. 210 °C; **2:** R_1 = CH_3, R_2 = n-C_3H_7, m.p. 151 °C;
3: R_1 = CH_3, R_2 = n-C_5H_{11}, m.p. 168 °C; **4:** R_1,R_2 = C_2H_5, m.p. 150 °C;
5: R_1 = CH_3, R_2 = CF_3, m.p. 190 °C; **6:** R_1 = CH_3, R_2 = H, m.p. 159 °C;
7: R_1 = CH_3, R_2 = Ph, m.p. 109 °C; **8:** R_1 = CH_3, R_2 = CH_2Ph, m.p. 175 °C;
6: R_1 = H, R_2 = H, m.p. 160 °C.

Figure 5.65 Synthesis of furan-aliphatic and furan-aromatic dihydryzides.

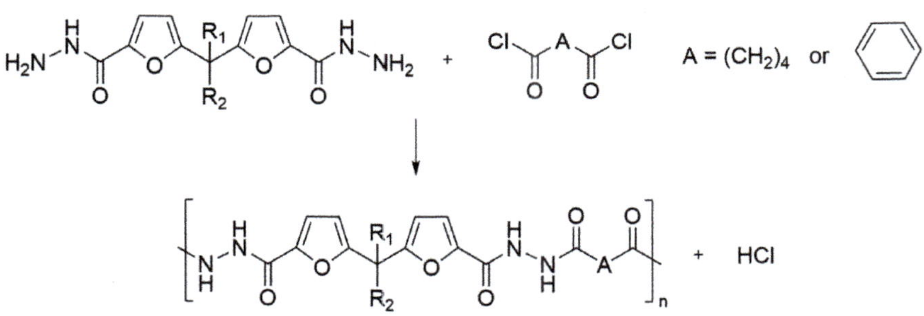

Figure 5.66 Synthesis of a series of furan polyhydryzides.

The furan-aromatic polymer was also converted into the corresponding poly-1,3,4-oxadiazole by cyclodehydration with thionyl chloride in 95% yield and the new furan polymer structure verified spectroscopically [206].

5.5 Conjugated Furan Oligomers and Polymers

The interest in π-conjugated oligomers and polymers has been growing steadily in the last fifty years since the pioneering work on poly(acetylene), because of the vast implications related to the promising use of these materials in various electronic and optical developing technologies. The field has therefore become a hot area of research and development [207]. Figure 5.67 shows a selection of the most important polymers in the area, all of which have received and continue to receive much attention. The relevant features here cover a wide scope which includes, among others, semiconducting, electrochromism, photoluminescence, energy storage, sensors, coatings, and electronic devices such as organic field-effect transistors, photovoltaic cells, and light-emitting devices, with the notable advantage of flexible fiber and film-forming aptitude [207].

The irruption of furan-based materials is more recent and covers essentially the two types of structures discussed here. It is important to explain that the conjugated structures arising from both the polycondensation of FA (Figure 3.17) and the cationic polymerization of some 2-alkenylfurans (Figure 4.19) described above are not suitable for the practical applications highlighted here, because of their irregular and irreproducible characteristics.

The first report on the purported polymerization of 5-methylfurfural (5MF) described its treatment in a strongly basic medium which produced a black insoluble material whose characterization was regrettably too poor to establish a

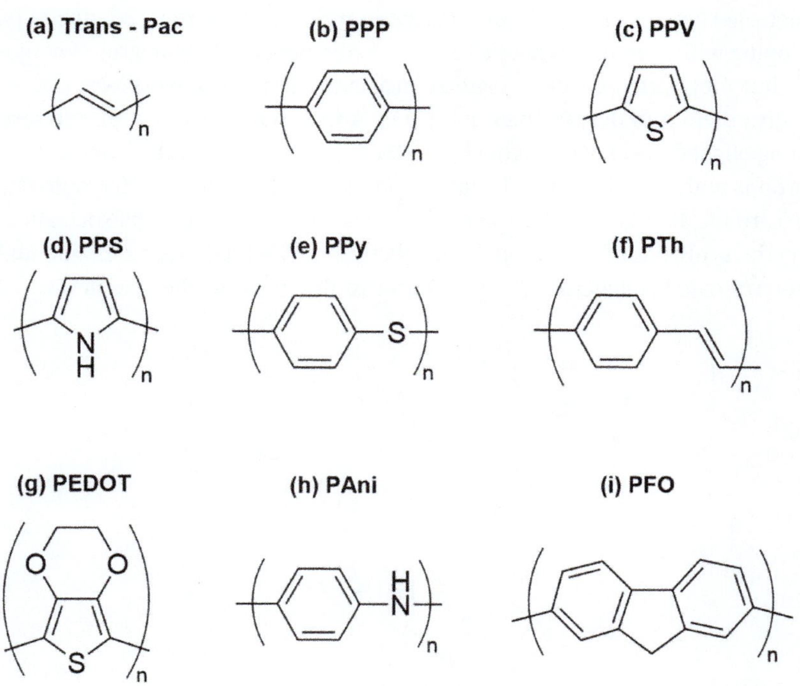

Figure 5.67 A selection of the most relevant conjugated polymers excluding the two furan homologues discussed here.

macromolecular structure [208]. A decade later, a brief report described the nucleophilic aldol-crotonic self-condensation of MF to give its dimer and furylene-vinylene low-DP polymers [209]. A detailed study was then published, in which tBuOK was found to promote both dimerization and polymerization, but with an important caveat concerning the mechanism and hence the required experimental procedure [210]. The polycondensation step in this process can be represented by the mechanism shown in Figure 5.68, but in the actual system, the dimer formed in the first step does not react with itself because the unsaturation between the two furan rings inhibits the reactivity of the methyl group [210], i.e., under classical polycondensation conditions, the process stops at that point. For the growth to continue, it is therefore necessary to continuously feed the medium with FA, since its methyl group can now react with any terminal aldehyde function, irrespective of the reached DPn. The ensuing polymer shown in Figure 5.68 is first conjugated furan polymer not included in Figure 5.67.

The 5MF dimer was fully characterized and became an interesting reagent *per se*, thanks to its photochemical behavior [210], as further discussed below. As for the actual polymers, the optimization of their synthetic conditions [210,211] gave deep-red materials (PVF) with DPn values as high as 100 with a regular structure, stability in air and good solubility in chlorinated solvents, but no Tg or melting were detected because of their high cohesive energy. Their electrical conductivity, after doping with iodine, was about 1 S cm^{-1} confirming the high degree of conjugation, but they suffered from a brittle consistency and hence work was carried out to turn them into flexible materials [211]. A number of potential plasticizers, including oligoethers of Mn ranging from 600 to 2,000 Da, were mixed in different proportions with PVF, but in all instances phase separation marred the outcome of these trials. It was therefore decided to attempt a chemical plasticization, namely the synthesis of block copolymers between PVF's aldehyde function and amino-terminated oligoethers. After successfully testing the possibility of

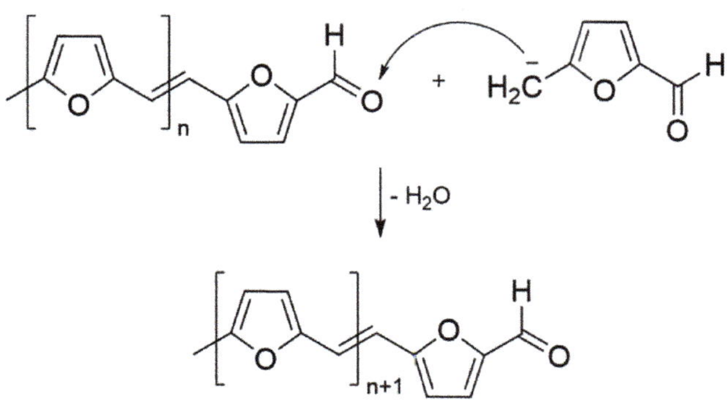

Figure 5.68 The growth mechanism in the nucleophilic polycondensation of 5MF. For the first step (dimerization), n = 0.

preparing these Schiff bases by reacting the model 5MF dimer, PVF was submitted to the same treatment and gave products with glass transition temperatures not exceeding −38 °C, i.e., materials which were too soft for any mechanical performance that would allow spinning fibers or casting films. In other words, the situation went from an intractable polymer to one that lacked sufficient cohesion, hence a relative failure. A more efficient photochemical approach was therefore later applied, as discussed below.

A comprehensive study was also conducted in order to prepare well-defined MF oligomers [212] with n = 1, 2 and 3, and a 4–8 mixture (Figure 5.68), which were thoroughly characterized. In particular, the low energy peak in their electronic spectra was respectively at 387, 445, 490 and 510 nm, suggesting that the molecules were planar only up to the tetramer, then partly losing the expected incremental conjugation. After protonation with HSO_3CF_3, these maxima moved to 527, 615, 755, and 900 nm, respectively as a result of the charge delocalization [212]. These oligomers displayed both photochemical activity (discussed below for the dimer) and photoluminescence, as shown in Figure 5.69, where both mercury-arc and laser-light excitation produced emissions at wavelengths that increased with n as revealed by the colors in Figure 5.69. This behavior was also found to occur when light excitation was replaced by applying a potential field, giving electroluminescence with the same features.

A recent contribution to furylene–vinylene oligomers [213] did not provide any important improvement because on the one hand, their syntheses are more cumbersome and less "green" than that using MF [210–212] and on the other hand, the lack of terminal aldehyde function reduces considerably their further use as the precursor to novel materials. As for their characterization, the study again deals with electronic spectra and only adds oxidation potentials [213].

The photodimerization of the 5MF dimer has been discussed earlier, as shown in Figure 2.11, where a clean single [π2 + π2] cyclodimerization was fully established [214]. Its application to photocrosslinking operations was exploited in a

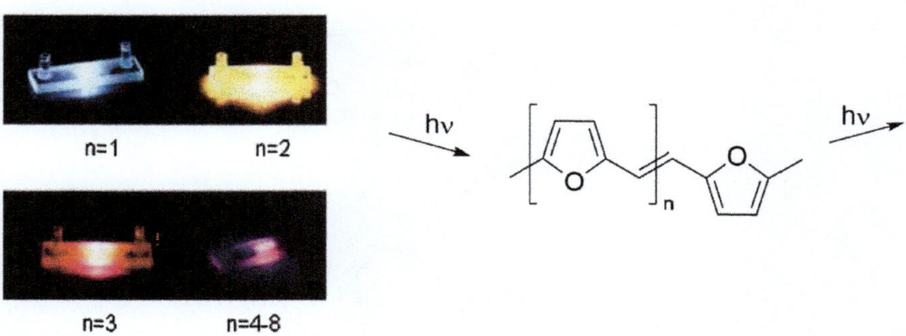

Figure 5.69 The photoluminescent behavior of furylene-vinylene oligomers.

number of investigations, starting with a study of polymer electrolytes networks prepared by a photochemical process [214]. Commercial mono-, di- and tri-aminated oligoethers were made to react with the 5MF dimer through the Schiff condensation reactions and the ensuing furan-terminated products submitted to UV to induce the dimer cyclocoupling, as sketched in Figure 5.70. A single coupling gave a linear structure comprising a central 5MF dimer and two oligoether tails, whereas the photochemical interaction of an oligoether terminated by two 5MF dimer units produced a linear polymer incorporating alternating motifs of flexible oligoether chains and cyclized 5MF dimers. Finally, the use of a branched molecule bearing three terminal 5MF moieties generated a crosslinked structure that was ideally suited to be used as polymer electrolytes in such devices as solid-state batteries [215].

A similar strategy was applied to a different type of polymer electrolyte [216], in which chitosan was modified by two branching reactions, one introducing

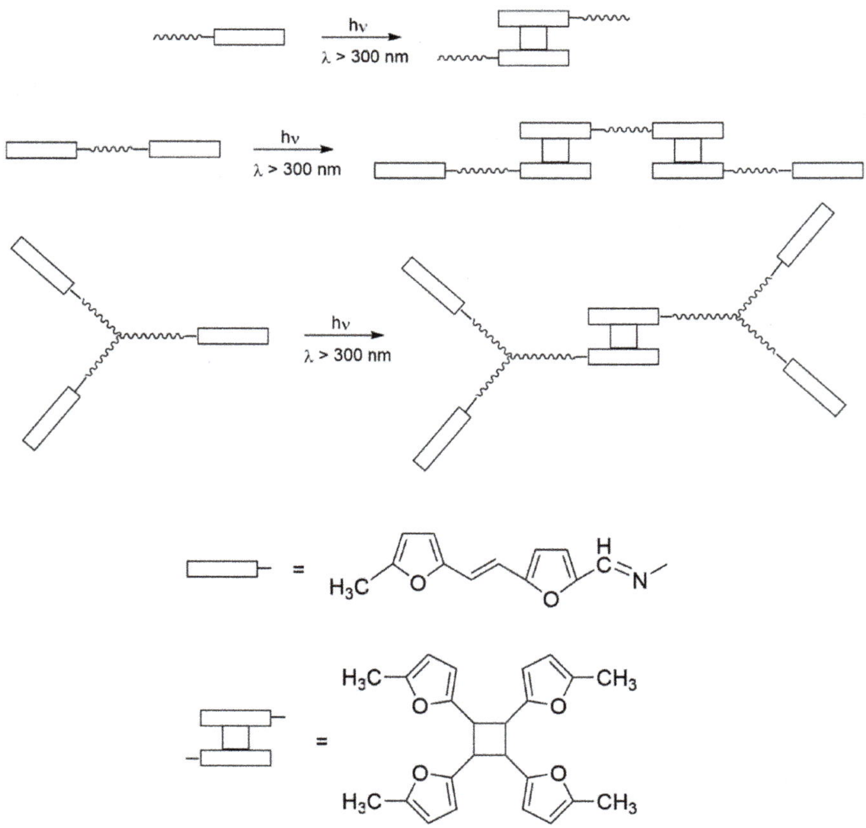

Figure 5.70 Photochemical synthesis of oligoether-based polymer electrolytes with different macromolecular architectures. The wavy lines represent the oligoether chains.

polyether strands and the other appending 5MF dimer moieties, the former in large excess with respect to the latter. The ensuing material was cast as thin films before being submitted to a photochemical crosslinking involving the furan chromophore, as shown in Figure 5.71. Both these polyelectrolyte networks gave high ionic conduction in the presence of $LiClO_4$ and had the advantage of a fast and efficient photochemical crosslinking.

Another type of technical materials was also investigated using this same approach, namely the possibility of photocrosslinking polymer films used in printing plates [217]. Commercial poly(vinyl alcohol) (PVA) was partly acetalyzed (5 to 20 unit %) by the 5MF dimer, as shown in Figure 5.72, following a procedure

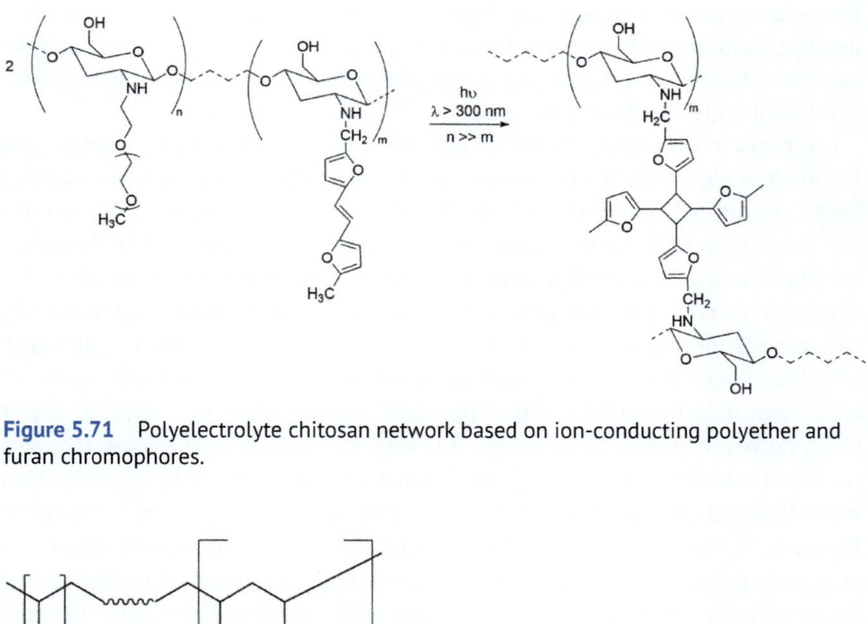

Figure 5.71 Polyelectrolyte chitosan network based on ion-conducting polyether and furan chromophores.

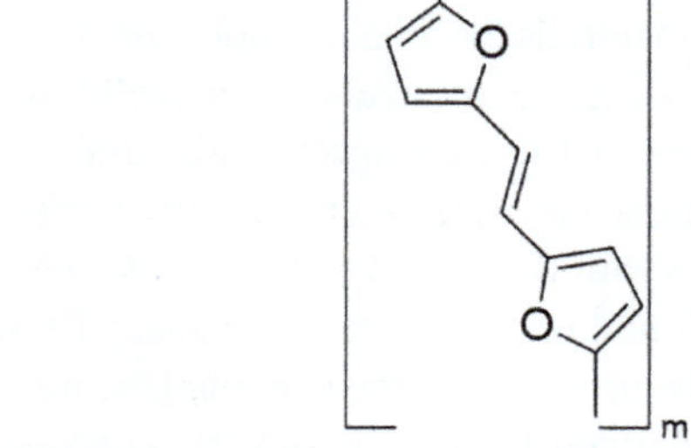

Figure 5.72 PVA acetalyzed with the 5MF dimer. Here n»m.

previously optimized with PVA and furfural [218], and the modified polymer spread as a thin film on an aluminum plate before submitting it to near-UV irradiation, which induced its rapid crosslinking through the photocoupling mechanism shown in Figures 5.70 and 5.71. This study proved technically viable for developing competitive negative printing plates with respect to commercial counterparts [219].

It must be emphasized that the photodimerization of the 5MF dimer is a reversible process, whose return to the starting conjugated structure can be activated by either far-UV irradiation, or by heat, a well-documented process for cyclobutanes generated photochemically. This is an important feature in terms of the possibility of recycling the different crosslinked polymers discussed above.

It can be concluded that 5MF, a cheap renewable compound industrially available as a side product in the generation of furfural, is a very useful building block for the preparation of conjugated oligomers and polymers which display semiconducting, luminescent, and photochemical properties with different potential applications, such as polymer reversible photocrosslinking or polymers for electronic and optical technologies.

Conjugated oligo- and poly-furans (Figure 5.73 shows their ideal structure, not included in Figure 5.67), the second type of conjugated furan polymers discussed here, have attracted attention for the last 40 years [220], apart from the cationic polymerization of furan discussed earlier, which failed to produce such regular structures and only gave oligomeric materials of limited potential interest.

Polyfurans (PFu), mostly prepared by electrochemical processes, but also by cationic polymerization, have been the subject of a thorough review in 2008 [221], which dealt with preparation, structural characterization, and properties of a variety of materials purported to reflect the ideal structure shown in Figure 5.73. It is clear, however, that in all instances this was not attained in the sense that side reactions interfered with such perfect growth, as pointed out by some authors who identified the frequent presence of irregular units, some of which as given in Figure 5.74. Thus, if on the one hand these polymers displayed some promising electronic properties for conductivity and electrochemical behavior, they lacked on the other hand the performances and reliability exhibited by most of the conjugated polymers shown in Figure 5.68. The numerous studies reviewed [221] are not discussed here because these drawbacks somehow limited the scope of the general topic in terms of the viability of the materials for reliable applications.

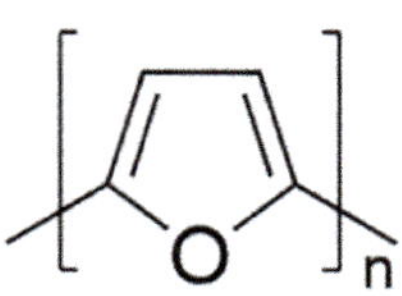

Figure 5.73 The monomer unit in the simplest oligo- and poly-furans.

A radical change took place in 2010 when a qualitative breakthrough was reported in a comprehensive study of the synthesis and characterization of structurally perfect individual oligofurans with DPs of 5 to 9 [222]. These powdery materials, with melting temperatures ranging from 209 to 328 °C from the pentamer to the octamer (the nonamer was vacuum-sublimed at 265 °C with partial decomposition), were soluble in some organic solvents, stable in air up to 250 °C and their electric spectra indicated a progressive bathochromic shift of the maximum absorption of their DP from 388 to 430 nm following the increase in conjugation. Likewise, their double-peak fluorescence spectra gave the same trend from 449 to 507 nm for their lower energy maximum. This behavior is entirely similar to that of the 5MF oligomers discussed above (see Figure 5.69). Their synthesis was carried out by Stille coupling reactions between tributyltin oligofurans **1** and **2** and dibromoolifurans **3** and **4** (Figure 5.75), the furan dimer and trimer being already well known.

The same group further showed that these materials had a clear potential to be exploited as organic "green" semiconductors [223] and other optoelectronic materials [224] as good candidates to replace polythiophene homologues, with the remarkable advantage of their renewable origin. This research program was pursued with another major contribution in which different oligofurans were shown to polymerize electrochemically to give, for the first time, polymers with regular

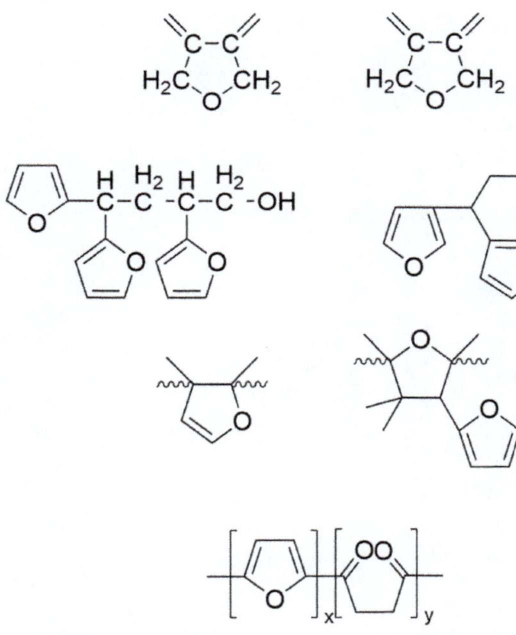

Figure 5.74 Anomalous monomer units detected in differently synthesized polyFu.

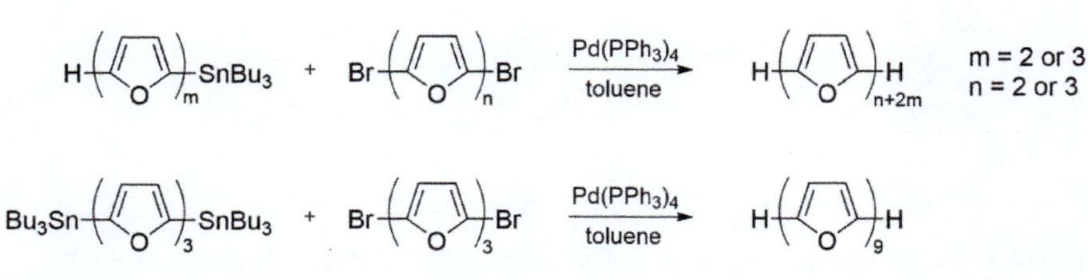

Figure 5.75 The first reported synthesis of well defined "long" oligofurans.

structures [225]. Figure 5.76 shows the three oligomers synthesized for this study using the same procedure as for the oligomers depicted in Figure 5.75 [222], and Figure 5.77 their corresponding polymers. Thanks to the use of oligomeric monomers instead of furan, milder conditions applied for the electrochemical

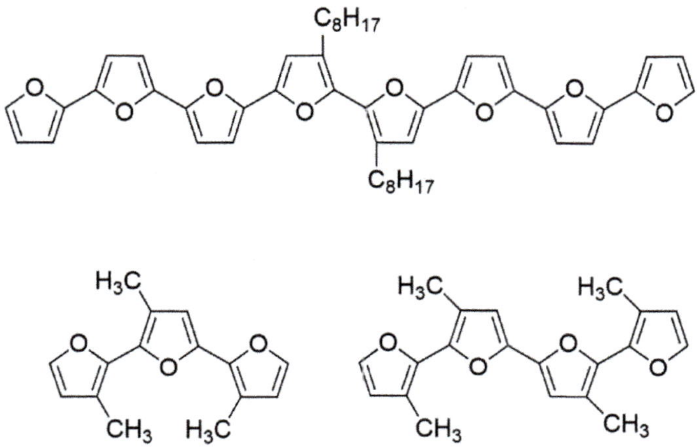

Figure 5.76 Oligofurans bearing C3 or C4 alkyl substituents.

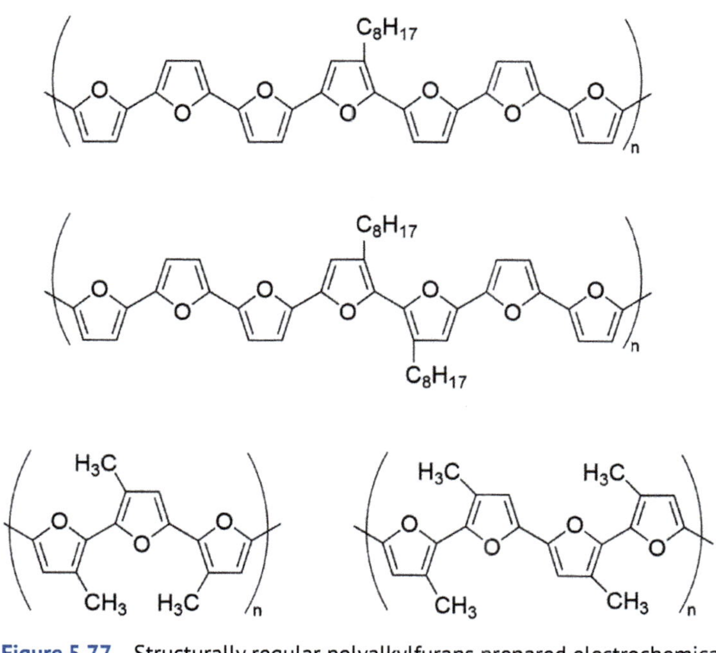

Figure 5.77 Structurally regular polyalkylfurans prepared electrochemically from the monomers shown in Figure 5.76.

deposition of the polymer film on the electrode (mostly lower applied potentials) insured the avoidance of the perturbing side reactions previously encountered in the numerous experiments involving furan and discussed above [221], i.e., these new PFus had a regular sequence of repeat units.

Although molecular weights were not reported, the free-standing films were otherwise fully characterized in terms of electronic conductance (up to 1 S cm^{-1} after doping), electrochemical stability, and electronic spectroscopy. Interestingly, the alkyl substitutions on the rings increased the flexibility of the polymers because of the reduced macromolecular packing density, hence their higher flexibility associated with the corresponding decrease in their Tg compared with the unsubstituted PFu [225].

Interesting progress on this topic followed with studies dealing with the preparation of the helical photostable ester-substituted PFu shown in Figure 5.78 [226]. A living-type polymerization mechanism insured narrow molecular weight distributions with Mn values reaching 10 kDa. To the best of our knowledge, this is the first report of a molecular weight for a PFu. Replacing the alkyl groups with ester moieties drastically reduced the effect of singlet-oxygen-induced photodegradation thanks to the electron-withdrawing nature of the ester substituents. In a subsequent paper [227], this group investigated the role of regiochemistry on the specific structure of the polymer and reached the conclusion that its actual "monomer unit" was the alternating dimeric conformation given in Figure 5.79.

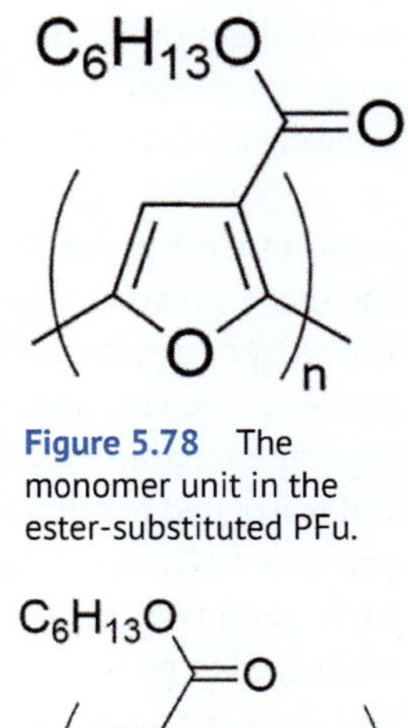

Figure 5.78 The monomer unit in the ester-substituted PFu.

In a different vein, a number of oligo- and poly-Fu incorporating variable contents of boron atoms in their chains as borane units, were synthesized via Stille coupling in order to study the role of the dopant element on the properties of the materials [228]. The presence of boron in the main chain was found to enhance the stability of the polymers to oxidation and reduce their electronic excitation and photoemission energy to an extent that was a function of boron content in a given macromolecule, thus providing the possibility of color tuning the materials.

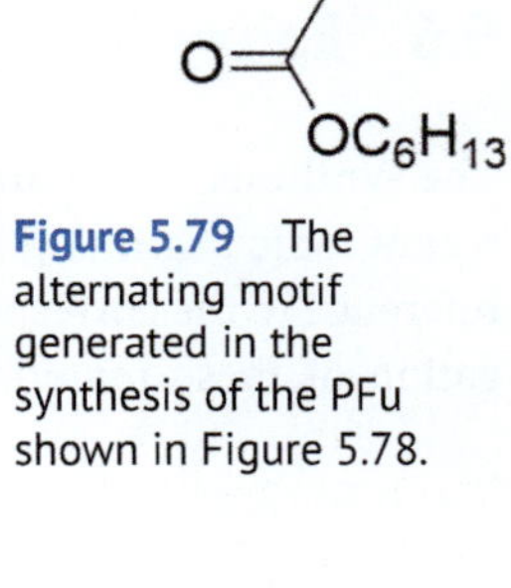

Figure 5.79 The alternating motif generated in the synthesis of the PFu shown in Figure 5.78.

In conclusion, this section on conjugated furan oligomers and polymers heralded the arrival of a novel family of materials for optoelectronic and photochemical applications, with performances comparable to, or better than, the well-established thiophene counterparts, with the major added advantage of being derived from renewable resources.

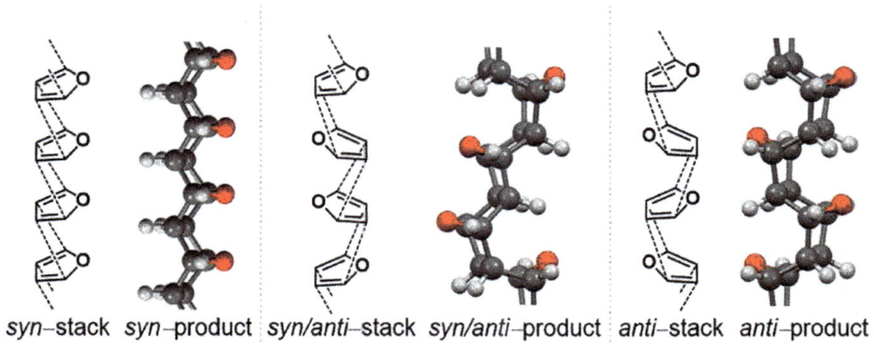

Figure 5.80 The structure of the furan nanothread. Reprinted with permission from Ref. [230a]. Copyright 2021 American Chemical Society.

A novel structure of polyfuran, namely furan nanothreads [229], has recently come to enrich this type of materials. Nanothreads are emerging one-dimensional sp^3-hybridized materials with high predicted tensile strength and a tunable band gap. They are prepared by compressing aromatic or nonaromatic small molecules to pressures of 15 to 30 GPa. Here, liquid furan was compressed at room temperature to 8–10 GPa under the illumination of a broad UV source, yielding a solid with a DPn of about 100. The lower degree of aromaticity of furan, compared with that of benzene or pyridine, allowed the use of a considerably more modest pressure to achieve the nanothread formation. This reduction of the onset pressure was shown to be due to an initial [4 + 4] cycloaddition, followed by a sequence of [4 + 2] cycloadditions associated with the polymerization process. The structure of such nanothreads is shown in Figure 5.80 [229]. This regular enchainment, observed during compression is partly lost when the material is isolated at ambient pressure because of lattice dilation. Despite the gain in pressure reduction with respect to aromatic counterparts, one is still dealing with high compression in terms of practical applications, and further research is envisaged to go further in this direction, including more specific photochemistry and the use of catalysts. A similar study has also recently described the preparation of FDCA nanothreads [230].

5.6 Epoxy Resins

The synthesis and characterization of furan-based epoxy polymers is a relatively recent topic beginning in the last decade, as recently reviewed, particularly with reference to the corresponding furan monomers [231]. The first thorough investigation of these materials [232] compares the properties of thermosets mostly

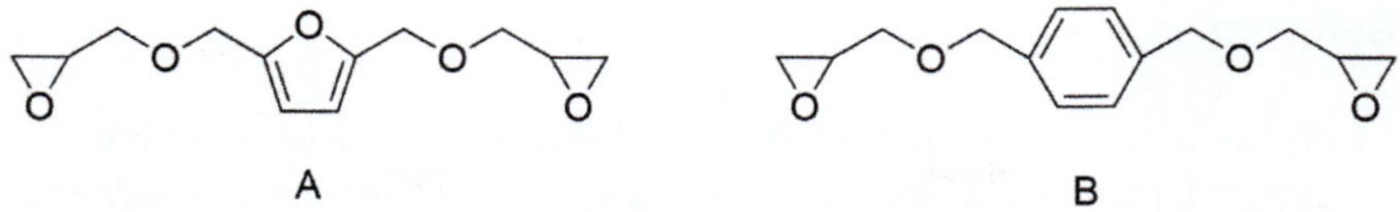

Figure 5.81 The diepoxy monomers used in a comparative study of their amine-based thermosets.

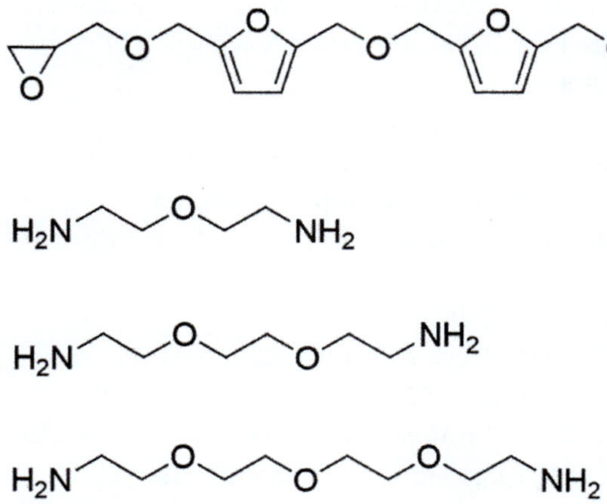

Figure 5.82 Epoxy resins based on a difuran diepoxy and three different diamines.

prepared from the furan diepoxy A and its aromatic counterpart B, shown in Figure 5.81, reacting with aliphatic and aromatic diamines. The furan network displayed higher Tg and storage modulus than those of the aromatic homologue, but the ether linkages in both monomers reduced these values compared with those of the more rigid classical phenolic glycidyl ethers.

In a second study by the same group [233], difuran diamines like those shown in Figure 5.51, were used in order to prepare fully biobased epoxy networks which displayed Tg values above 50 °C, storage moduli of 3.5 GPa and good thermal stability up to 270 °C in argon. Moreover, as already pointed out above, particularly in the context of furan-based polyurethanes, they produced a high char yield of about 40% at 800 °C, confirming this general tendency of furan-rich polymers.

A few years later, a novel difuran diepoxy monomer was synthesized and polymerized with oxoaliphatic diamines, as shown in Figure 5.82 [234]. The ensuing networks displayed a good fire-retardancy aptitude.

The two furan diepoxy monomers shown above were also combined with different aliphatic and aromatic diamines as curing agents giving cured networks [235] and their thermal and mechanical properties indicated that these materials could readily replace fossil-based counterparts.

References

1 (a) Ravve, A. (2012). *Principles of Polymer Chemistry*, 3e. Springer, New York, USA, 801. (b) Rudin, A. and Choi, P. (2013). *The Elements of Polymer Science and Engineering*, 3e. Elsevier, Waltham, USA, 563 p. (c) Cowie, J.M.G. and Arrighi, V. (2007). *Polymers: Chemistry and Physics of Modern Materials*, 3e. CRC Press, Boca Raton, USA, 499 p.

2 Nurmi, L., Jaatinen, S., Saleem, F., et al. (2018). From Biomass to Value-added Furan-based Platform Chemicals. FURCHEM and CatBio Roadmap. *VTT Technology* 326. ISBN 978-951-38-8633-2.

3 Zhang, Q., Song, M., Xu, Y., et al. (2021). *Prog. Polym. Sci.* 120: 101430.

4 Pang, J., Zheng, M., Sun, R., et al. (2016). *Green Chem.* 18: 342–359.

5 The Coca-Cola Company. (2023). Available online: https://www.coca-colacompany.com/news/100-percent-plant-based-plastic-bottle. Accessed January 11th, 2023.

6 Gandini, A. (1977). *Adv. Polym. Sci.* 25: 47–96.

7 Gandini, A. and Belgacem, M.N. (1997). *Progr. Polym. Sci.* 22: 1203–1379.

8 Moreau, C., Belgacem, M.N., and Gandini, A. (2004). *Top. Catal.* 27: 11–30.

9 Gandini, A. and Belgacem, M.N. (2008). In: *Monomers, Polymers and Composites from Renewable Resources*. (eds. M.N. Belgacem and A. Gandini). Elsevier, Amsterdam, The Netherlands.

10 Gandini, A. (2008). *Macromolecules* 41: 9491–9504.

11 Gandini, A. (2010). *Polym. Chem.* 1: 245–251.

12 Gandini, A. (2011). *Green Chem.* 13: 1061–1083.

13 Gandini, A. and Lacerda, T.M. (2015). *Prog. Polym. Sci.* 48: 1–39.

14 Bui, T.V., Crossley, S., and Resasco, D.E. (2016). In: *Chemicals and Fuels from Bio-Based Building Blocks*. (eds. F. Cavani, S. Albonetti, F. Basile, and A. Gandini). Wiley-VCH, Weinheim.

15 Gandini, A., Lacerda, T.M., Carvalho, A.J.F., and Trovatti, E. (2016). *Chem. Rev.* 116: 1637–1669.

16 Gandini, A. and Lacerda, T.M. (2022). In: *Macromolecular Engineering: From Precise Synthesis to Macroscopic Materials and Applications*, 2e. (eds. Y. Gnanou, N. Hadjichristidis, K. Matyjaszewski, M. Muthukumar, and S. Sheiko). John Wiley & Sons, Hoboken, NJ.

17 (a) Moore, J.A. and Kelly, J.E. (1978). *Macromolecules* 11: 568–573. (b) Moore, J.A. and Kelly, J.E. (1979). *Polymer* 20: 627–628. (c) Moore, J.A. and Kelly, J.E. (1978). *J. Polym. Sci., Polym. Chem. Ed.* 16: 2407–2409. (d) Moore, J.A. and Kelly, J.E. (1984). *J. Polym. Sci., Polym. Chem. Ed.* 22: 863–864. (e) Moore, J.A. and Partain, E.M. (1983). *Macromolecules* 16: 338–339.

18 (a) Khrouf, A., Boufi, S., El Gharbi, R., Belgacem, N.M., and Gandini, A. (1996). *Polym. Bull.* 37: 589–596. (b) Khrouj, A., Abid, M., Boufi, S., El Gharbi, R., and

Gandini, A. (1998). *Macromol. Chem. Phys.* 199: 2755–2765. (c) Chaabouni, A., Gharbi, S., Abid, M., et al. (1999). *J. Soc. Chim. Tun.* 4: 547–558. (d) Khrouj, A., Boufi, S., El Gharbi, R., and Gandini, A. (1999). *Polym. Int.* 48: 649–659. (e) Gharbi, S., Andreolety, J.-P., and Gandini, A. (2000). *Eur. Polym. J.* 36: 463–472.

19 (a) Gandini, A., Silvestre, A.J.D., Pascoal Neto, C., Sousa, A.F., and Gomes, M. (2009). *J. Polym. Sci. A Polym. Chem.* 47: 295–298. (b) Gomes, M., Gandini, A., Silvestre, A.J.D., and Reis, B. (2011). *J. Polym. Sci. A Polym. Chem.* 49: 3759–3768.

20 Huang, Y.-T., Wong, J.-J., Huang, C.-J., Li, C.-L., and Jang, G.-W.B. (2016). In: *Chemicals and Fuels from Bio-Based Building Blocks*. (eds. F. Cavani, S. Albonetti, F. Basile, and A. Gandini). Wiley-VCH, Weinheim.

21 Ram, V.J., Sethi, A., Nath, M., and Pratap, R. (2019). In: *The Chemistry of Heterocycles*. (eds. V.J. Ram, A. Sethi, M. Nath, and R. Pratap). Elsevier, Amsterdam, The Netherlands.

22 Abid, M., Triki, R., Abid, S., and El Gharbi, R. (2009). *J. Soc. Chim. Tun.* 11: 45–52.

23 Triki, R., Abid, M., Tessier, M., et al. (2013). *Eur. Polym. J.* 49: 1852–1860.

24 Hbaieb, S., Bougarech, A., Kammoun, W., and Abid, M. (2017). *J. Macromol. Sci. A* 54: 524–533.

25 Abid, M., Mhiri, S., Bougarech, A., Triki, R., and Abid, S. (2020). *Des. Monomers Polym.* 23: 16–24.

26 Abid, M., Aden Ali, M., Bernard, J., et al. (2014). *Polym. Sci. Ser B+* 56: 290–297.

27 Wang, L., Eguchi, Y., and Chen, E.Y.-X. (2017). *Ind. Eng. Chem. Res.* 56: 11380–11387.

28 Ali, D.K. and Sweileh, B.A. (2018). *J. Polym. Environ.* 26: 1940–1949.

29 Gaitonde, V., Lee, K., Kirschbaum, K., and Sucheck, S.J. (2014). *Tetrahedron Lett.* 55: 4141–4145.

30 Miyagawa, N., Ogura, T., Okano, K., et al. (2017). *Chem. Lett.* 46: 1535–1538.

31 Kainulainen, T.P., Sirviö, J.A., Sethi, J., Hukka, T.I., and Heiskanen, J.P. (2018). *Macromolecules* 51: 1822–1829.

32 Kainulainen, T.P., Hukka, T.I., Özeren, H.D., et al. (2020). *Biomacromolecules* 21: 743–752.

33 Amarasekara, A.S., Nguyen, L.H., Okorie, N.C., and Jamal, S.M. (2017). *Green Chem.* 19: 1570–1575.

34 Ahmed, A.M., Kainulainen, T.P., Sirviö, J.A., and Heiskanen, J.P. (2022). *Biomacromolecules* 23: 1803–1811.

35 (a) Yi, J., Xu, Z., Wu, Y., et al. (2022). *ACS Sustainable Chem. Eng.* 10: 4404–4414. (b) Xu, Z., Yan, P., Liu, K., et al.(2016). *ChemSusChem* 9: 1255–1258.

36 Warlin, N., Gonzalez, M.N.G., Mankar, S., et al. (2019). *Green Chem.* 21: 6667–6684.

37 (a) Mou, Z. and Chen, E.Y.-X. (2016). *ACS Sustainable Chem. Eng.* 4: 7118–7129. (b) Zang, H. and Chen, E.Y.X. (2015). *Int. J. Mol. Sci.* 16: 7143–7158.

38 Wilson, J.F. and Chen, E.Y.-X. (2019). *ACS Sustainable Chem. Eng.* 7: 7035–7046.

39 Kainulainen, T.P., Parviainen, T.A.O., Sirviö, J.A., McGeachie, L.J.R., and Heiskanen, J.P. (2023). *ACS Macro Lett.* 12: 147–151.

40 Sousa, A.F., Vilela, C., Fonseca, A.C., et al. (2015). *Polym. Chem.* 6: 5961–5983.

41 Matsuda, K., Matsuda, H., Horie, H., and Komuro, T. (2007) *Polymer Compound and Method of Synthesizing the Same.* PCT Patent WO2007/052847 A1, filed November 6, 2006 and issued May 10, 2007.

42 Kato, S. and Kasai, A. (2008). *Method for Producing Polyester Resin Including Furan Structure.* Japanese Patent JP2008/291244, filed April 24, 2007, issued December 04, 2008.

43 Gomes, F.W., Lima, R.C., Piombini, C.R., et al. (2018). *Macromol. Symp.* 381: 1800129.

44 Knopp, R.J.I., Vogelzang, W., van Haveren, J., and van Es, D.S. (2013). *J. Polym. Sci., Part A: Polym. Chem.* 51: 4191–4199.

45 Jiang, M., Liu, Q., Zhang, Q., Ye, C., and Zhou, G. (2012). *J. Polym. Sci. Polym. Chem.* 50: 1026–1036.

46 (a) Gruter, G.-J.M., Sipos, L., and Dam, M.A. (2012). *Comb. Chem. High Thr. Scr.* 15: 180–188. (b) Ma, J., Pang, Y., Wang, M., et al. (2012). *J. Mater. Chem.* 22: 3457–3461. (c) Papageorgiou, G.Z., Tsanaktsis, V., and Bikiaris, D.N. (2014). *Phys. Chem. Chem. Phys.* 16: 7946–7958. (d) Gopalakrishnan, P., Narayan-Sarathy, S., Ghosh, T., Mahajan, K., and Belgacem, M.N. (2014). *J. Polym. Res.* 21: 340. (e) Thiyagarajan, S., Vogelzang, W., Knoop, R.J.I., et al. (2014). *Green Chem.* 16: 1957–1966.

47 Celli, A., Colonna, M., Gandini, A., et al. (2016). In: *Chemical and Fuels from Bio-Based Building Blocks*, Vol. 1. (eds. F. Cavani, S. Albonetti, F. Basile, and A. Gandini). Wiley-VCH Verlag GmbH & Co, Weinheim, Germany, 315–350.

48 Matos, M., Sousa, A.F., Fonseca, A.C., et al. (2014). *Macromol. Chem. Phys.* 215: 2175–2184.

49 Gruter, G.-J. (2015). European Green Chemistry in Action: PEF, a 100% Bio-based Polyester: Synthesis, Properties & Sustainability. In: *EuroNanoForum (2015)*, Riga, Latvia June 10–12, 2015.

50 Burgess, S.K., Leisen, J.E., Kraftschik, B.E., et al. (2014). *Macromolecules* 47: 1383–1391.

51 Loos, K., Zhang, R., Pereira, I., et al. (2020). *Front. Chem.* 8: DOI: 10.3389/fchem.2020.00585.

52 de Jong, E., Dam, M.A., Sipos, L., and Gruter, G.-J.M. (2012). In: *Biobased Monomers, Polymers and Materials*, ACS Symposium Series (eds. P.B. Smith and R.A. Gross). American Chemical Society, Washington, DC, 1–13.

53 Burgess, S.K., Mikkilineni, D.S., Yu, D.B., et al. (2014). *Polymer* 55: 6870–6882.

54 Burgess, S.K., Mikkilineni, D.S., Yu, D.B., et al. (2014). *Polymer* 55: 6861–6869.

55 (a) De Jong, E., Visser, H.A., Dias, A.S., Harvey, C., and Gruter, G.-J.M. (2022). *Polymers* 14: 943. (b) Avantium. (2021). The Journey of Avantium's PEF towards Commercialization: History, Present and Future. Available online: https://www.

avantium.com/wp-content/uploads/2021/12/20211209-The-Journey-of-PEF-towards-commercialisaton-history-present-and-future.pdf. Accessed January 31st, 2023.

56 Rosenboom, J.-G., Hohl, D.K., Fleckenstein, P., Storti, G., and Morbidelli, M. (2018). *Nat. Commun.* 9: 2701.

57 Fei, X., Wang, J., Zhang, X., et al. (2022). *Polymers* 14: 625.

58 Gandini, A. and Lacerda, T.M. (2022). *Macromol. Mater. Eng.* 307: 2100902.

59 (a) Hu, L., Xu, J., Zhou, S., et al. (2018). *ACS Catal.* 8: 2959–2980. (b) Silvianti, F., Maniar, D., Boetje, L. and Loos, K. (2020). In: *Sustainability & Green Polymer Chemistry*. Vol. 2 Biocatalysis and Biobased Polymers, ACS Symposium Series eBooks (eds. H.N. Cheng and R. Gross). Oxford University Press, Oxford, UK.

60 Guidotti, G., Soccio, M., García-Gutiérrez, M.C., et al. (2020). *ACS Sustainable Chem. Eng.* 8 (25): 9558–9568.

61 Guidotti, G., Soccio, M., García-Gutiérrez, M.-C., et al. (2019). *ACS Sustainable Chem. Eng.* 7 (21): 17863–17871.

62 Tsanaktsis, V., Terzopoulou, Z., Nerantzaki, M., Papageorgiou, G.Z., and Bikiaris, D.N. (2016). *Mater. Lett.* 178: 64–67.

63 Papageorgiou, D.G., Guigo, N., Tsanaktsis, V., et al. (2016). *Ind. Eng. Chem. Res.* 55: 5315–5326.

64 Soares, M.J., Dannecker, P.-K., Vilela, C., et al. (2017). *Eur. Polym. J.* 90: 301–311.

65 Todea, A., Bîtcan, I., Aparaschivei, D., et al. (2019). *Polymers* 11: 1402.

66 Gubbels, E., Jasinska-Walc, L., and Koning, C.E. (2013). *J. Polym. Sci. A Polym. Chem.* 51: 890–898.

67 Thiyagarajan, S., Vogelzang, W., Knoop, R.J.I., et al. (2014). *Green Chem.* 16: 1957–1966.

68 Tsanaktsis, V., Terzopoulou, Z., Exarhopoulos, S., et al. (2015). *Polym. Chem.* 6: 8284–8296.

69 Arnaud, S.P., Wu, L., Chang, M.-A.W., et al. (2017). *Faraday Discuss.* 202: 61–77.

70 Honeycutt, D.S., Charbonneau, W.F., North, A.J., et al. (2022). *Polymer* 242: 124584.

71 Terzopoulou, Z., Kasmi, N., Tsanaktsis, V., et al. (2017). *Materials* 10: 801.

72 Tian, S., Cao, X., Luo, K., et al. (2021). *Biomacromolecules* 22: 4823–4832.

73 Rhein, M., Demharter, A., Felker, B., and Meier, M.A.R. (2022). *ACS Appl. Polym. Mater.* 4: 6514–6520.

74 Bianchi, E., Soccio, M., Siracusa, V., et al. (2021). *ACS Sustainable Chem. Eng.* 9: 11937–11949.

75 Wu, J., Eduard, P., Thiyagarajan, S., et al. (2015). *ChemSusChem* 8: 67–72.

76 Shahni, R.K., Mabin, M., Wang, Z., et al. (2020). *Polym. Chem.* 11: 6081–6090.

77 Xie, H., Wu, L., Li, B.-G., and Dubois, P. (2019). *Biomacromolecules* 20: 353–364.

78 Xie, H., Wu, L., Li, B.-G., and Dubois, P. (2018). *Ind. Eng. Chem. Res.* 57: 13094–13102.

79 Meng, H., Li, Z., Wu, L., et al. (2022). *J. Appl. Polym. Sci.* 139: e51921.

80 Zhang, X., Wang, J., Dong, Y., Wang, Q., and Zhu, J. (2022). *J. Appl. Polym. Sci.* 139: e52627.
81 Balla, E.D., Papadopoulos, L., Ainali, N.M., et al. (2022). *Eur. Polym. J.* 176: 111429.
82 Zhu, Y., Wang, J., Mahmud, S., et al. (2022). *Eur. Polym. J.* 181: 111654.
83 Sousa, A.F., Matos, M., Freire, C.S.R., Silvestre, A.J.D., and Coelho, J.F.J. (2013). *Polymer* 54: 513–519.
84 Xie, H., Wu, L., Li, B.-G., and Dubois, P. (2020). *Eur. Polym. J* 134: 109856.
85 Li, J., Tu, Y., Lu, H., et al. (2021). *Polymer* 237: 124313.
86 Jia, Z., Wang, J., Sun, L., et al. (2018). *J. Appl. Polym. Sci.* 136: 47291.
87 (a) Kwiatkowska, M., Kowalczyk, I., Kwiatkowski, K., Szymczyk, A., and Rosłaniec, Z. (2016). *Polymer* 99: 503–512. (b) Kwiatkowska, M., Kowalczyk, I., Szymczyk, A., and Gorący, K. (2018). *Polimery* 63: 594–602.
88 Suriano, R., Gonzalez, M.N.G., and Turri, S. (2021). *J. Polym. Environ.* 29: 1122–1133.
89 Sousa, A.F., Fonseca, A.C., Serra, A.C., et al. (2016). *Polym. Chem.* 7: 1049–1058.
90 Dai, J., Ma, S., Teng, N., et al. (2017). *Ind. Eng. Chem. Res.* 56: 2650–2657.
91 Hofmann, M.A., Shahid, A.T., Garrido, M., et al. (2022). *ACS Sustainable Chem. Eng.* 10: 3442–3454.
92 Lomelí-Rodriguez, M., Corpas-Martínez, J.R., Willis, S., Mulholland, R., and Lopez-Sanchez, A. (2018). *Polymers* 10: 600.
93 Molares-Huerta, J.C., Martínez de Ilarduya, A., León, S., and Muñoz-Guerra, S. (2018). *Macromolecules* 51: 3340–3350.
94 Wang, X., Wang, Q., Liu, S., and Wang, G. (2019). *Eur. Polym. J.* 115: 70–75.
95 Ouyang, Q., Liu, J., Li, C., et al. (2019). *Polym. Chem.* 10: 5594–5601.
96 (a) Shen, A., Wang, G., Wang, J., et al. (2021). *Eur. Polym. J.* 147: 110317. (b) Fei, X., Zhu, Y., Wang, J., Jia, Z., and Liu, X. (2022). *Polym. Adv. Technol.* 33: 2265–2275.
97 (a) Zhang, H., Jiang, M.M., Wu, Y., et al. (2021). *Green Chem.* 23: 2347–2448. (b) Zhang, H., Jiang, M., Wu, Y., et al. (2021). *ACS Sustainable Chem. Eng.* 9: 6799–6809.
98 Kwiatkowska, M., Kowalczyk, I., Kwiatkowski, K., Szymczyk, A., and Jędrzejewski, R. (2017). *Polymer* 130: 26–38.
99 Kwiatkowska, M., Kowalczyk, I., Kwiatkowski, K., and Zubkiewicz, A. (2020). *Polymers* 12: 271.
100 Zhang, X., Ma, S., Shen, A., et al. (2020). *J. Appl. Polym. Sci.* 137: e49365.
101 Feng, Y., Li, Y., Ye, X., et al. (2022). *J. Mater. Sci.* 57: 10997–11012.
102 Wang, G., Hou, M., Jiang, M., et al. (2020). *Polym. Degrad. Stabil.* 179: 109244.
103 Wang, G., Dong, Y., Hao, X., Zhang, L., and Sun, R. (2023). *Polym. Bull.* 80: 5373–5395.
104 Lv, X., Luo, F., Zheng, L., et al. (2022). *J. Appl. Polym. Sci.* 139: e53122.

105 Du, W.-Q., Fu, T., Li, X.-L., et al. (2022). *ACS Sustainable Chem. Eng.* 10: 14240–14247.
106 Soccio, M., Costa, M., Lotti, N., et al. (2016). *Eur. Polym. J.* 81: 397–412.
107 Gubbels, E., Jasinka-Walc, L., Noordover, B.A.J., and Koning, C.E. (2013). *Eur. Polym. J.* 49: 3188–3198.
108 Kasmi, N., Ainali, N.M., Agapiou, E., et al. (2019). *Polym. Degrad. Stabil.* 169: 108983.
109 Geng, Y., Wang, Z., Hu, X., et al. (2019). *Eur. Polym. J.* 114: 476–484.
110 Guidotti, G., Soccio, M., Lotti, N., et al. (2019). *Polym. Degrad. Stabil.* 169: 108963.
111 Wang, G., Jiang, M., Zhang, Q., et al. (2018). *Polym. Degrad. Stabil.* 153: 292–297.
112 Paszkiewicz, S., Irska, I., Zubkiewicz, A., et al. (2021). *Polymers* 13: 397.
113 Feng, S., Jiang, Z., and Qiu, Z. (2023). *Polymers* 15: 85.
114 Aparaschivei, D., Todea, A., Frissen, A.E., et al. (2019). *Pure Appl. Chem.* 91: 397–408.
115 Wang, G., Jiang, M., Zhang, Q., et al. (2018). *Polym. Degrad. Stabil.* 153: 272–280.
116 Cheng, Z., Cheng, J., Chen, J., et al. (2021). *J. Wuhan Univ. Technol.-Mat. Sci. Edit.* 36: 557–561.
117 Kluge, M., Papadopoulos, L., Magaziotis, A., et al. (2020). *ACS Sustainable Chem. Eng.* 8: 10812–10821.
118 Min, J., Tingting, L., Qiang, Z., Ying, C., and Guangyuan, Z. (2015). *J. Renew. Mater.* 3: 120–141.
119 Bouyahya, C., Patrício, R., Paço, A., et al. (2022). *Polymers* 14: 3868.
120 Kasmi, N., Terzopoulou, Z., Chebbi, Y., et al. (2022). *Polym. Degrad. Stabil.* 195: 109804.
121 Wang, J., Zhang, X., Fei, X., et al. (2022). *ACS Sustainable Chem. Eng.* 10: 13595–13606.
122 Wei, P., Li, L., Wang, L., et al. (2023). *Eur. Polym. J.* 183: 111738.
123 Zaidi, S., Thiyagarajan, S., Bougarech, A., et al. (2019). *Polym. Chem.* 10: 5324.
124 (a) Zubkiewicz, A., Irska, I., Walkowiak, K., Dryzek, J., and Paszkiewicz, S. (2022). *Express Polym. Lett.* 16: 1099–1112. (b) Walkowiak, K., Irska, I., Zubkiewicz, A., Dryzek, J., and Paszkiewicz, S. (2022). *Polymers* 14: 2295. (c) Zubkiewicz, A., Irska, I., Miadlicki, P., et al. (2021). *J. Mater. Sci.* 56: 19296–19309. (d) Paszkiewicz, S., Walkowiak, K., Irska, I., et al. (2023). *Polymer* 269: 125740.
125 Zeng, C., Seino, H., Ren, J., Hatanaka, K., and Yoshie, N. (2013). *Macromolecules* 46: 1794–1802.
126 Ikezaki, T., Matsuoka, R., Hatanaka, K., and Yoshie, N. (2014). *J. Polym. Sci. Part A: Polym. Chem.* 52: 216–222.
127 Yoshie, N., Yoshida, S., and Matsuoka, K. (2019). *Polym. Degrad. Stabil.* 161: 13e18.
128 Zhang, Y., Li, T., Xie, Z., et al. (2017). *Ind. Eng. Chem. Res.* 56: 3937–3946.

129 Ragno, D., Di Carmine, G., Brandolese, A., et al. (2019). *Chem. Eur. J.* 25: 14701.

130 Guillaume, L., Marshall, A., Niessen, N., Ni, P., Gauvin, R.M., and Thomas, C.M. (2021). *Green Chem.* 23: 6931–6935.

131 (a) Zhang, J., Liu, Y., Qi, Z., He, L., and Peng, L. (2020). *BioResources* 15: 4502–4527. (b) Kainulainen, T.P., Hukka, T.I., Özeren, H.D., et al. (2020). *Biomacromolecules* 21: 743–752. (c) Fei, X., Wang, J., Zhu, J., Wang, X., and Liu, X. (2020). *ACS Sustainable Chem. Eng.* 8: 8471–8485.

132 Jiang, Y., Woortman, A.J.J., van Ekenstein, G.O.R.A., Petrovic, D.M., and Loos, K. (2014). *Biomacromolecules* 15: 2482–2493.

133 Jiang, Y., Woortman, A.J.J., van Ekenstein, G.O.R.A., and Loos, K. (2015). *Polym. Chem.* 6: 5198–5211.

134 Maniar, D., Jiang, Y., Woortman, A.J.J., van Dijken, J.,, and Loos, K. (2019). *ChemSusChem* 12: 990–999.

135 Pellis, A., Weinberger, S., Gigli, M., Guebitz, G.M., and Farmer, T.J. (2020). *Eur. Polym. J.* 130: 109680.

136 (a) Lasseuguette, E., Gandini, A., Belgacem, M.N., and Timpe, H.-J. (2005). *Polymer* 46: 5476–5483. (b) Lasseuguette, E., Gandini, A., and Timpe, H.-J. (2005). *J. Protoch. Photobio. A* 174: 222–228.

137 Anslyn, E.V. and Dougherty, D.A. (2006). Strain and Stability. In: *Modern Physical Organic Chemistry*. University Science Books, Sausaltio, California, USA, Chapter 2, 65–144.

138 Wang, Z., Miller, B., Mabin, M., et al. (2017). *Sci. Rep.-UK* 7: 13704.

139 Wang, Z., Kastern, B., Randazzo, K., et al. (2015). *Green Chem.* 17: 4720–4724.

140 Takeuchi, D. (2013). In: *Stereoregular Polymers, Encyclopedia of Polymer Science and Technology*. (ed. H.F. Mark). John Wiley & Sons, Inc., New York, Vol. 4, 254–286. DOI 10.1002/0471440264.pst595.

141 Wang, Z.D., Elliott, Q., Wang, Z., et al. (2018). *ACS Sustainable Chem. Eng.* 6: 8136–8141.

142 Wang, Z., Scheuring, M., Mabin, M., et al. (2020). *ACS Sustainable Chem. Eng.* 8: 8909–8917.

143 Siahkali, M.J.M., Guigo, N., and Sbirrazzuoli, N. (2023). *Macromolecules* 56: 404–415.

144 Wang, J., Wang, Y., Liu, B., and Fu, T. (2022). *New J. Chem.* 46: 16781.

145 Pezzana, L., Melilli, G., Delliere, P., et al. (2022). *Prog. Org. Coat.* 173: 107203.

146 Cureton, L.T., Napadensky, E., Annunziato, C., and La Scala, J.J. (2017). *J. Appl. Polym. Sci.* 134: 45514.

147 Huang, W., Hu, X., Zhai, J., Zhu, N., and Guo, K. (2020). *Mater. Today Sustain.* 10: 100049.

148 Gharbi, S. and Gandini, A. (1999). *Acta Polym.* 50: 293–297.

149 Gharbi, S., Afli, A., El Gharbi, R., and Gandini, A. (2001). *Polym. Int.* 50: 509–514.

150 Abid, M., El Gharbi, R., and Gandini, A. (2004). *Polymer* 45: 5793–5801.

151 Miyagawa, N., Suzuki, T., Okano, K., et al. (2018). *J. Polym. Sci. A Polym. Chem.* 56: 1516–1519.

152 Yeh, I.-C., Rinderspacher, B.C., Andzelm, J.W., Cureton, L.T., and La Scala, J. (2014). *Polymer* 55: 166–174.

153 Luo, K., Wang, Y., Yu, J., Zhu, J., and Hu, Z. (2016). *RSC Adv.* 6: 87013–87020.

154 Cao, M., Zhang, C., He, B., Huang, M., and Jiang, S. (2017). *Macromol. Res.* 25: 722–729.

155 Yu, Z., Deng, R., Rao, G., et al. (2020). *J. Polym. Sci.* 58: 2140–2150.

156 Cao, K., Liu, Y., Yuan, F., et al. (2021). *High Perform. Polym.* 33: 1083–1092.

157 Mao, L., Pan, L., Ma, B., and He, Y. (2022). *J. Polym. Environ.* 30: 1072–1079.

158 Kamran, M., Davidson, M.G., Tsanaktsis, V., van Berkel, S., and de Vos, S. (2022). *Eur. Polym. J.* 178: 111496.

159 Kamran, M., Davidson, M.G., de Vos, S., Tsanaktsis, V., and Yeniad, B. (2022). *Polym. Chem.* 13: 3433–3443.

160 Ai, T., Zou, G., Feng, W., et al. (2022). *ACS Sustain. Chem. Eng.* 10: 13792–13804.

161 Xie, S., Yang, J., Wang, X., and Yang, J. (2022). *Eur. Polym. J.* 162: 110932.

162 Xie, S., Wang, W., Yao, J., et al. (2023). *J. Polym. Environ.* DOI: 10.1007/s10924-023-02828-8.

163 Jiang, Y., Maniar, D., Woortman, A.J.J., Alberda van Ekenstein, G.O.R., and Loos, K. (2015). *Biomacromolecules* 16: 3674–3685.

164 Jiang, Y., Maniar, D., Woortman, A.J.J., and Loos, K. (2016). *RSC Adv.* 6: 67941–67953.

165 Maniar, D., Hohmann, K.F., Jiang, Y., et al. (2018). *ACS Omega* 3: 7077–7085.

166 (a) Mitiakoudis, A., Gandini, A., and Cheradame, H. (1985). *Polym. Commun.* 26: 246–249. (b) Mitiakoudis, A. and Gandini, A. (1991). *Macromolecules* 24: 830–835.

167 Gaddour, M., Bougarech, A., Abid, M., and Abid, S. (2019). *J. Polym. Res.* 26: 74.

168 Yang, M.-L., Wu, Y.-X., Liu, Y., Qiu, -J.-J., and Liu, C.-M. (2019). *Polym. Chem.* 10: 6217–6226.

169 Mohagheghnezhad, M. and Rafiee, Z. (2020). *Polym. Bull.* 77: 4739–4757.

170 Ali, D.K. (2022). *J. Polym. Environ.* 30: 4102–4113.

171 Abid, M., El Gharbi, R., and Gandini, A. (2000). *Polymer* 41: 3555–3560.

172 Lankenau, A.W. and Kanan, M.W. (2020). *Chem. Sci.* 11: 248–252.

173 Woroch, C.P., Cox, I.W., and Kanan, M.W. (2023). *J. Am. Chem. Soc.* 145: 697–705.

174 Wang, X., Liu, C., Xing, Z., et al. (2022). *Macromolecules* 55: 8857–8865.

175 Sun, J., Qian, L., and Li, J. (2020). *Polymer* 210: 122994.

176 Patil, Y., Mahindrakar, J., Ankushrao, S., et al. (2021). *Int. J. Polym. Anal. Ch.* 26: 707–720.

177 Liu, Y.-L. and Chen, Y.-W. (2007). *Macromol. Chem. Phys.* 208: 224–232.

178 Luo, K., Li, J., Duan, G., et al. (2018). *Polymer* 142: 33–42.

179 Zhao, J., Chen, S., Zhao, J., Zhang, Z., and Zhang, J. (2021). *J. Polym. Res.* 28: 101.

180 Chen, S., Feng, Y., Zhang, Z.-Y., et al. (2022). *Polym. Eng. Sci.* 62: 1269–1280.
181 Luo, K., Song, G., Wang, Y., et al. (2019). *ACS Appl. Polym. Mater.* 1: 944–952.
182 Quillerou, J., Belgacem, M.N., Gandini, A., Rivero, J., and Roux, G. (1989). *Polym. Bull.* 21: 555–562.
183 Belgacem, M.N., Quillerou, J., Gandini, A., Rivero, J., and Roux, G. (1989). *Eur. Polym. J.* 25: 1125–1130.
184 Belgacem, M.N., Quillerou, J., and Gandini, A. (1993). *Eur. Polym. J.* 29: 1217–1224.
185 Moss, E.K. (1982). *J. Cell. Plastics* 19: 240–244.
186 (a) Cawse, J.L., Stanford, J.L., and Still, R.H. (1984). *Makromol. Chem.* 185: 697–708. (b) Cawse, J.L., Stanford, J.L., and Still, R.H. (1984). *Makromol. Chem.* 185: 709–723. (c) Cawse, J.L., Stanford, J.L., and Still, R.H. (1985). *Br. Polym. J.* 17: 233–238.
187 Boufi, S., Belgacem, M.N., Quillerou, J., and Gandini, A. (1993). *Macromolecules* 26: 6706–6717.
188 Boufi, S., Gandini, A., and Belgacem, M.N. (1995). *Polymer* 36: 1689–1696.
189 Gandini, A. and Belgacem, M.N. (2022). Furan Resins. In: *Handbook of Thermoset Plastics*, 4e (ed. A. Dodiuk). Elsevier, Oxford, UK, 83–96.
190 Neumann, C.N.D., Bulach, W.D., Rehahn, M., and Klein, R. (2011). *Macromol. Rapid Comm.* 32: 1373–1378.
191 Lee, J., Baek, S., Moon, H.H., Son, S.U., and Song, C. (2021). *ACS Appl. Polym. Mater.* 3: 5767–5777.
192 Hui, Z. and Gandini, A. (1991). *Polym. Bull.* 26: 383–390.
193 Abid, S., Matoussi, S., El Gharbi, R., and Gandini, A. (2006). *Polym. Bull.* 57: 43–50.
194 Holfinger, M.S., Conner, A.H., Holm, D.R., and Hill, C.G. (1995). *J. Org. Chem.* 60: 1595–1598.
195 Trovatti, E., Gonçalves, I.G., Carvalho, A.J.F., and Gandini, A. (2020). *Polym. Intern.* 69: 688–692.
196 (a) Li, S.Q., Zhao, J.B., Zhang, Z.Y., Zhang, J.Y., and Yang, W.T. (2015). *RSC Adv.* 5: 6843–6852. (b) Ma, S., Liu, C., Sablong, R.J., et al. (2016). *ACS Catal.* 6: 6883–6891. (c) Jiang, S., Cheng, H.Y., Shi, R.H., et al. (2019). *ACS Appl. Mater. Interfaces* 11: 47413–47421. (d) Wang, P.X., Ma, X.Y., Li, Q.H., et al. (2016). *RSC Adv.* 6: 54013–54019. (e) White, B.T., Migliore, J.M., Mapesa, E.U., et al. (2020). *RSC Adv.* 10: 18760–18768. (e) Kébir, N., Benoit, M., Legrand, C., and Burel, F. (2017). *Eur. Polym. J.* 96: 87–96. (f) Dennis, J.M., Steinberg, L.I., Pekkanen, A.M., et al. (2018). *Green Chem.* 20: 243–249.
197 Jiang, Y., Yang, S., and Pan, X. (2022). *ACS Appl. Polym. Mater.* 4: 2197–2204.
198 Kumar, B.S. and Kumar, K.S.S. (2022). Polybenzoxazines: Recent Advances. In: *Handbook of Thermoset Plastics*, 4e (ed. A. Dodiuk). Elsevier, Oxford, UK, Ch. 3, 41–64.

199 (a) Lu, G., Dai, J., Liu, J., et al. (2020). *ACS Omega* 5: 3763–3773. (b) Mohamed, M.G., Li, C-J., Khan, M.A.R., et al. (2022). *Macromolecules* 55: 3106–3115.

200 (a) Liu, Y.L. and Chou, C.I. (2005). *J. Polym. Sci. Polym. Chem.* 43: 5267–5282. (b) Wang, C., Sun, J., Liu, X., Sudo, A., and Endo, T. (2012). *Green Chem.* 14: 2799–2806. (c) Wang, C., Zhao C., Sun, J., et al. (2013). *J. Polym. Sci. Polym. Chem.* 51: 2016–2023.

201 Yang, R., Han, M., and Hao, B. (2020). *Eur. Polym. J.* 131: 109706.

202 (a) Shen, X., Dai, J., Liu, Y., and Zhu, J. (2017). *Polymer* 122: 258–269. (b) Zhu, H.-X., Liu, Y., Wu, Y.-X., Qiu, J.-J., and Liu, C.-M. (2020). *React. Funct. Polym.* 155: 104716.

203 Hui, Z. and Gandini, A. (1992). *Eur. Polym. J.* 28: 1461–1469.

204 Méalares, C. and Gandini, A. (1996). *Polym. Int.* 40: 33–39.

205 (a) Xiang, T., Liu, X., Yi, P., et al. (2013). *Polym. Int.* 62: 1517–1523. (b) Tachibana, Y., Hayashi, S., and Kasuya, K. (2018). *ACS Omega* 3: 5336–5345.

206 Afli, A., Gharbi, S., El Gharbi, R., Le Bigot, Y., and Gandini, A. (2002). *Eur. Polym. J.* 38: 667–673.

207 (a) Reynolds, J.R., Thompson, B.C., and Skotheim, T.A. (Eds.) (2019). *Conjugated Polymers. Properties, Processing and Applications*. CRC Press, Boca Raton. (b) Lu, H., Li, X., and Lei, Q. (2021). *Front. Chem.* 9: 732132.

208 Kossmehl, G. and Yaridjanian, A. (1981). *Makromol. Chem.* 182: 3419–3426.

209 Gandini, A., Salon, M.C., Choura, M., et al. (1992). *Makromol. Chem. Macromol. Symp.* 60: 165–176.

210 (a) Méalares, C. and Gandini, A. (1994). *Trends Polym. Sci.* 2: 127–131. (b) Méalares, C., Hui, Z., and Gandini, A. (1996). *Polymer* 37: 2273–2279.

211 Méalares, C. and Gandini, A. (1996). *Eur. Polym. J.* 32: 1269–1272.

212 (a) Coutterez, C. and Gandini, A. (1998). *Polymer* 39: 7009–7014. (b) Coutterez, C., Goussé, C., Gheneim, R., Waig Fang, S., and Gandini, A. (2001). *ACS Symp. Ser.* 784: 98–109.

213 Mallet, C. and Frère, P. (2022). *Eur. J. Org. Chem.* e202201276. DOI: 10.1002/ejoc.202201276.

214 Lasseguette, E., Gandini, A., and Timpe, H.-J. (2005). *J. Photochem. Photobiol. A Chem.* 174: 222–228.

215 Albertin, L., Stagnaro, P., Coutterez, C., Le Nest, J.-F., and Gandini, A. (1998). *Polymer* 39: 6187–6189.

216 Velazquez-Morales, P., Le Nest, J.-F., and Gandini, A. (1998). *Electrochim. Acta* 43: 1275–1279.

217 Fang, S.W., Timpe, H.-J., and Gandini, A. (2002). *Polymer* 43: 3505–3510.

218 Goussé, C. and Gandini, A. (1997). *Eur. Polym. J.* 33: 667–671.

219 Gandini, A., Fang, S.W., Timpe, H.-J., and Baumann, H. (2001). *US Patent 6,270,938 B1 to the KODAK-POLYCHROME Co.*

220 Gandini, A. and Belgacem, M.N. (1997). *Progr. Polym. Sci.* 22: 1325–1327.

221 Gonzalez-Tejera, M.J., Sanchez de la Blanca, E., and Carrillo, I. (2008). *Synthetic Met.* 158: 165–189.

222 Gidron, O., Diskin-Posner, Y., and Bendikov, M. (2010). *J. Am. Chem. Soc.* 132: 2148–2150.

223 Gidron, O., Dadvand, A., Sheynin, Y., Bendikov, M., and Perepichka, D.F. (2011). *Chem. Commun.* 47: 1976–1978.

224 Gidron, O., Dadvand, A., Sun, E.W.-H., et al. (2013). *J. Mater. Chem. C* 1: 4358–4367.

225 Sheberla, D., Patra, S., Wijsboom, Y.H., et al. (2015). *Chem. Sci.* 6: 360–371.

226 Varni, A.J., Fortney, A., Baker, M.A., et al. (2019). *J. Am. Chem. Soc.* 141: 8858–8867.

227 Varni, A.J., Kawakami, M., Kramer, S.N., et al. (2022). *J. Polym. Sci.* 60: 579–589.

228 Fritze, L., Fest, M., Helbig, A., et al. (2021). *Macromolecules* 54: 7653–7665.

229 (a) Huss, S., Wu, S., Chen, B., et al. (2021). *ACS Nano* 15: 4134–4143. (b) Oburn, S.M., Huss, S., Cox, J., et al. (2022). *J. Am. Chem. Soc.* 144: 22026–22034.

230 Wang, X., Yang, X., Wang, Y., et al. (2022). *J. Am. Chem. Soc.* 144: 21837–21842.

231 Eid, M., Amenduri, B., and Boutevin, B. (2021). *ACS Sustainable Chem. Eng.* 9: 8018–8031.

232 Hu, F., La Scala, J.J., Sadler, J.M., and Palmese, G.R. (2014). *Macromolecules* 47: 3332–3342.

233 Hu, F., Yadav, S.K., La Scala, J.J., Sadler, J.M., and Palmese, G.R. (2015). *Macromol. Chem. Phys.* 216: 1441–1446.

234 Meng, J., Zeng, Y., Zhu, G., et al. (2019). *Polym. Chem.* 10: 2370–2375.

235 Jiang, Y., Yun, J., and Pan, X. (2022). *ACS Sustainable Chem. Eng.* 10: 16555–16562.

6

The Furan/Maleimide Diels–Alder Reaction Applied to Polymer Synthesis and Modification

The chemical principles associated with the Diels–Alder (DA) click reaction were briefly outlined in Section 2.5, with particular emphasis on the situations in which furan is the diene reagent. We now turn to the actual context of the macromolecular domains involving the intervention of the DA coupling and decoupling that implicate the furan/maleimide diene/dienophile combination. Two major areas are discussed here, namely (i) the synthesis of polymers in which multifunctional furans and maleimides are the reactive entities insuring the thermally reversible growth of linear, branched, and cross-linked macromolecules (i.e., step-growth systems), and (ii) the reversible cross-linking of linear or branched polymers incorporating furan (or maleimide) moieties induced by the intermolecular coupling with bismaleimides (or difurans). Schematically, the two types of interactions are shown in Figure 6.1, where F, M, and A stand, respectively, for a furan heterocycle, a maleimide group, and the corresponding DA adduct. Other more elaborate architectural constructs are of course possible, as indeed discussed in this section.

In all these reactions, the associated thermally reversible DA click coupling/decoupling is the same, as shown in Figure 6.2.

Two important considerations are necessary at this point. The first is that the *endo* and *exo* adducts, formed here in proportions that vary with the operating conditions, do not play any significant role on the properties of the various possible macromolecular structures schematized in Figure 6.1. The second has to do with the actual temperatures used in these processes in that for the coupling reaction, about 60 °C optimizes its rate, while insuring the maximum shift to the right of the equilibrium, whereas this is optimally shifted to the left (decoupling) above 100 °C, although it has been found that it should be limited to about 130 °C to avoid side reactions such as those involving the maleimide unsaturation.

As detailed in the first comprehensive review on the topic [1], studies of these polymers begun in the late 1980, but did not really take off before the end of the century, and a further acceleration has been occurring in the last decade with literary a proliferation of publications, many carrying original contributions, others

Furan Polymers and their Reactions, First Edition. Alessandro Gandini and Talita Martins Lacerda.

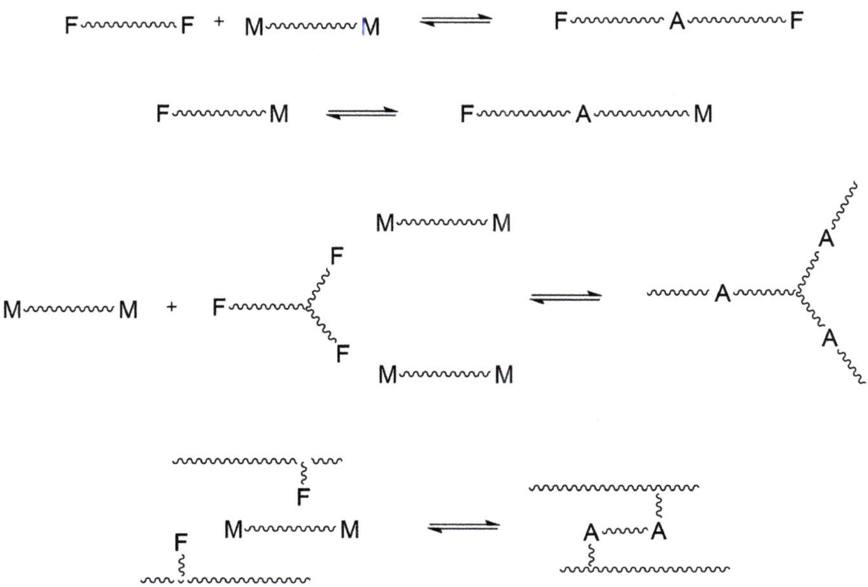

Figure 6.1 Schematic illustration of DA polycondensations and polymer cross-linking involving the thermally reversible coupling of furan (F) and maleimide (M) moieties to give the corresponding adduct (A).

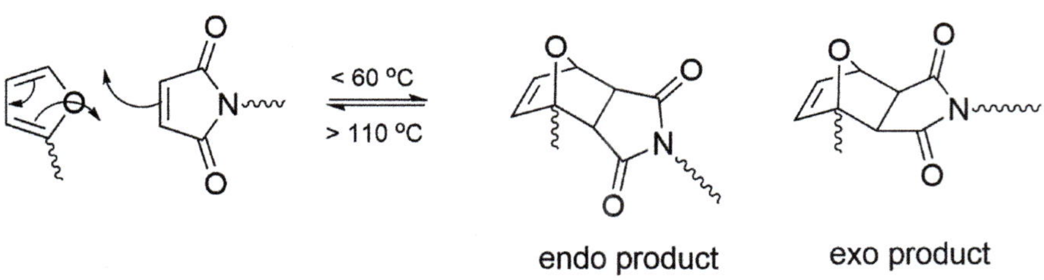

Figure 6.2 The mechanism of the thermally reversible furan/maleimide reaction.

being simple additions to well-established systems. Some other reviews on the application of the DA reaction to polymer science have also appeared [2]. The following survey covers what we consider the most relevant contributions to the area, with particular emphasis on recent qualitative advances.

6.1 Polycondensations (Step-growth Polymerizations)

Among several previous investigations [1, 2], the first *systematic* study of a linear AA + BB polymerization [3] examined a number of systems and covered their entire behavior, including monomers synthesis, polymerization kinetics, and the reversibility of the processes. Thus, for example, the linear polymerization shown in Figure 6.3 was followed by UV spectroscopy at different temperatures [3], taking advantage of the loss of conjugation associated with the conversion of the maleimide function to the corresponding adduct, to the detriment of the O = C - C = C - C = O moiety which gives a maximum absorption at 300 nm. Figure 6.4 displays the decrease of this peak as a function of polymerization time at 60 °C. The same technique demonstrated the occurrence of the reverse reaction carried out at 110 °C, as shown in Figure 6.5. These experiments, carried out at different temperatures, provided the kinetic parameters for the reaction, including its activation energy. The validity of the experimental approach was clearly confirmed by ^{1}H NMR studies. The polymerization/depolymerization cycle was repeated reproducibly several times [3].

Several similar investigations on AA + BB systems have followed this study, including a study of the relative proportion of *endo* and *exo* adduct conformers as a function of the operating conditions [4], one focusing on the dielectric properties of these linear DA polymers [5] and one based on different flexible difuran monomers and the stiff *N,N*-dimethyl-4-aminopyridine (DMAP) bismaleimide [6].

Few reports are available of AB-type DA linear polycondensation involving the furan/maleimide combination. To the best of our knowledge, the first such successful study was published in 2011 [7] using protected AB monomers to avoid

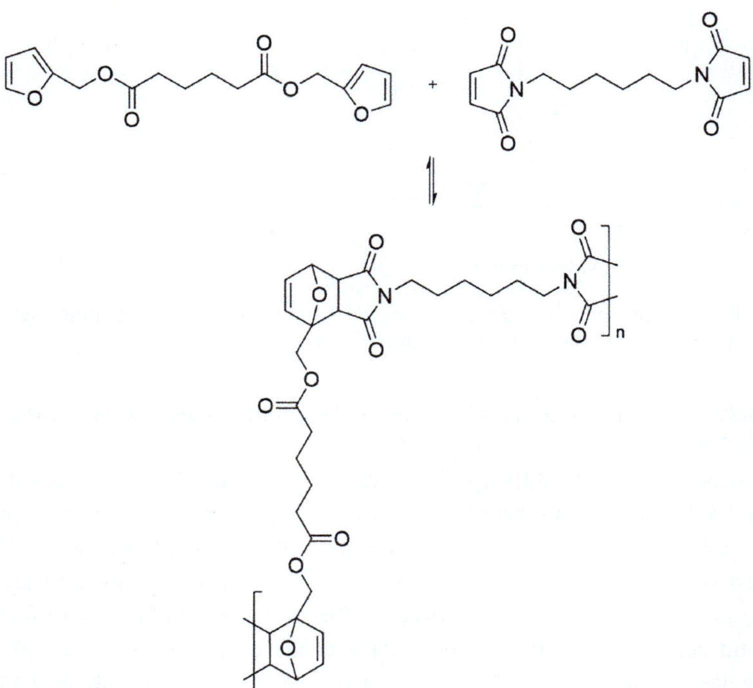

Figure 6.3 An example of studied linear DA polycondensation.

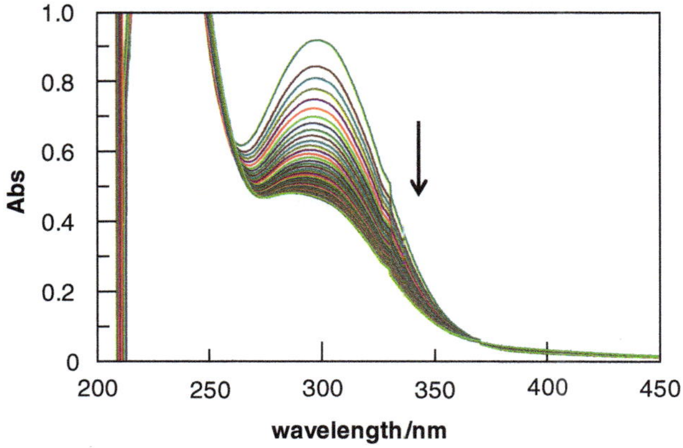

Figure 6.4 Progressive decrease in the 300 nm peak intensity at 60 °C as a function of polymerization time over 62 h, for the reaction in Figure 6.3.

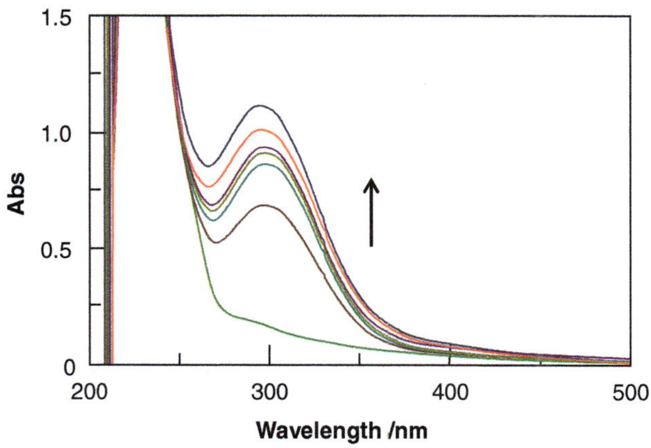

Figure 6.5 Progressive increase in the 300 nm peak intensity as a function of time over 24 h at 110 °C for the depolymerization reaction in Figure 6.3.

their premature polymerization, and decomposing the protecting adduct *in situ*, as shown in Figure 6.6.

A more recent study tackled the synthesis and polycondensation of the "shortest" possible AB monomer 2-furfurylmaleimide [8] using again the deprotection procedure, as shown in Figure 6.7. The reaction produced a solid polymer with the expected structure, a DPn of about 10 and a Tg of 90 °C, which however did not display any crystallinity probably because the monomer units were in fact both *exo* and *endo* conformers. The polymer was heated at 150 °C in a high-vacuum device provided with a liquid nitrogen cold finger, which allowed to release and collect the unprotected monomer, a crystalline solid (m.p. 92 °C), which did not repolymerize in the solid state, but did upon melting [8].

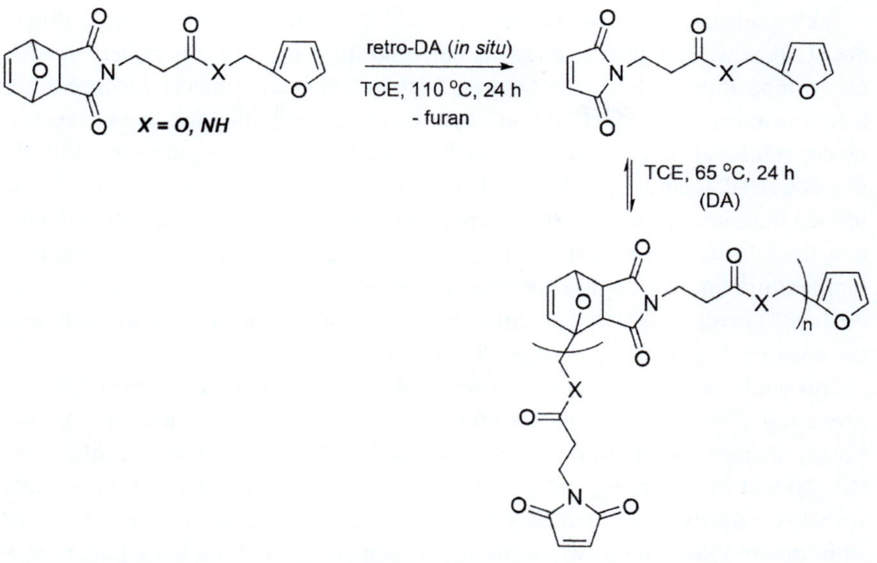

Figure 6.6 The DA polycondensation of AB furan/maleimide monomers.

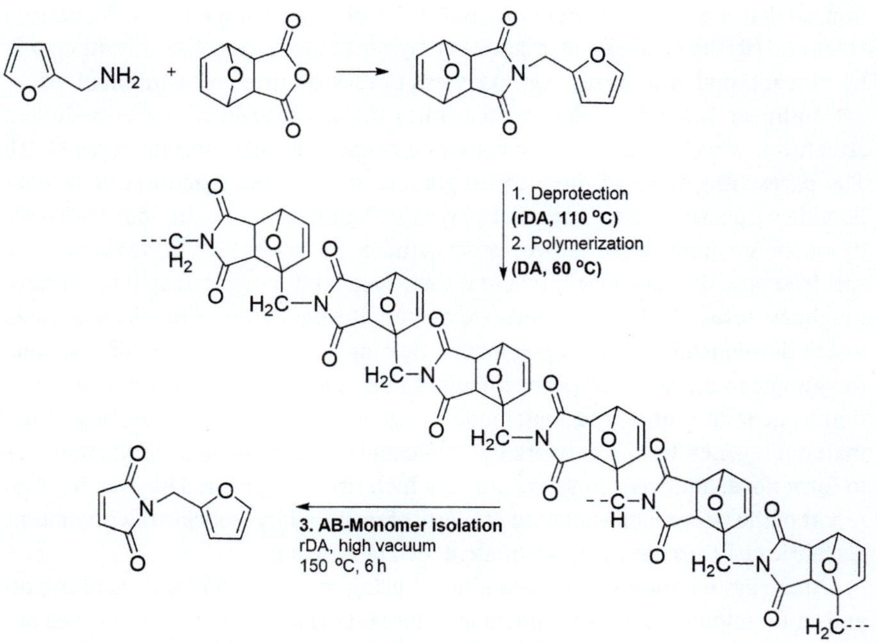

Figure 6.7 The DA system involving 2-furfurylmaleimide.

Taking advantage of the thermal reversibility of these systems, a very different strategy was adopted to show its potential for preparing copolymers of various compositions [9]. Two polyesters, based on the same diacid incorporating a furan/maleimide adduct and an aliphatic or an aromatic diol, were prepared by conventional polycondensation techniques as shown in Figure 6.8 and fully characterized, including their retro-DA mechanism and respective furan/maleimide AB monomer products. They were dissolved together in 1,1,2,2-tetrachloethane, the solution was heated at 110 °C for 45 min, cooled to 55 °C, and kept at that temperature for 24 h. Figure 6.9 gives a schematic view of the reactions involved during the heating (depolymerization at 110 °C) and cooling (copolymerization of the ensuing polymer fragments at 55 °C) periods.

This study provided a successful proof of concept related to the possibility of preparing copolymers from the rDA/DA cycle applied to a mixture of two homopolymers incorporating furan/maleimide DA adducts. These results open the way to a wider strategy based on the synthesis of a library of novel copolymers with two priority goals, namely (i) the possibility of fine-tuning their structure from random assemblies (when the rDA deconstruction of the homopolymers is carried out all the way to the F-M monomers), to blocky architectures with blocks of variable lengths (when the depolymerization is interrupted before its completion, so that the DA reconstruction involves reactive F-M oligomers of the desired size); and (ii) the synthesis of original copolymers which are not readily attainable by conventional procedures, e.g. epoxy-ester or amide-urethane combinations.

Non-linear DA polycondensation studies include branched and cross-linked structures as well as other more elaborate macromolecular architectures [1, 2]. The pacesetting starts of these investigations involved the reaction of a trismaleimide with a tetrafuran [10], as displayed in Figure 6.10. The thermal reversibility of the ensuing dense networks was studied in detail by ^{13}C NMR and UV spectroscopy, DSC and DMA, whereas their mendability was assessed by fracture toughness tests. The latter property stems from the possibility of repairing a crack, which developed in the solid polymer by heating it at 115–150 °C, while pressing the sample in the direction perpendicular to the crack, to allow the retro-DA reaction to generate furan and maleimide moieties. The subsequent cooling of the material regenerates the network by statistical DA recombination of the moieties to form novel adducts across the crack, which thus disappear. This was the first report of the later much exploited property of mendability associated with materials fabricated through the furan/maleimide DA reaction.

A more systematic study was conducted using various combinations of the di- and tri-functional monomers shown in Figure 6.11 [11]. The reactions, carried out using different initial furan/maleimide molar ratios, were followed by UV and NMR spectroscopy, as well as viscosity measurements. Three different [maleimide]/[furan] or [furan]/[maleimide] functional ratios *r*, viz. 1.0, 0.75, and 0.50, were applied for all the combinations and the relevant relationship providing the

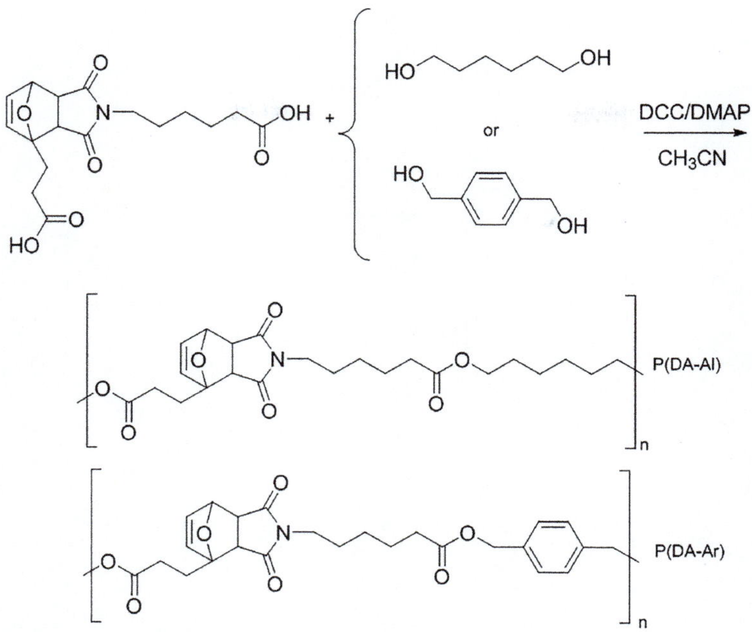

Figure 6.8 The aliphatic and the aromatic polyesters prepared for the DA copolymerization study.

critical yield *Pc*, associated with the inception of gel formation, was the *Flory–Stockmayer* equation, so that the corresponding *Pc* would be 71, 81, and 100% conversions. As expected, these reactions were accompanied by a progressive increase in viscosity, with formation of a gel (network) for molar rations r 1.0 and 0.75, whereas for $r = 0.5$, a highly viscous liquid was obtained at the end of the reactions. The fact that cross-linking did not occur with the latter systems was rationalized by considering that the gel point was predicted for 100% conversion *in the absence of cyclization reactions*, i.e., an ideal situation.

The depolymerizations, carried out at 110 °C and followed again via spectroscopic and viscometry analysis, proceeded as expected, albeit at slower rates than the corresponding linear polycondensations [3]. Again, various heating/cooling cycles were always found to reproduce the behavior of the first polymerization. The products isolated before the gel formation were analyzed by NMR spectroscopy and SEC. Evidence of highly branched structures was largely confirmed by the spectra and the very broad SEC tracings, which, however, also displayed sharp peaks corresponding to molecular weights of 5 to 10 kDa, suggesting the presence of large, cyclized macromolecules arising from self-condensation reactions, a well-known phenomenon in this type of non-linear polycondensations. Subsequent publications on these systems [2c, 2d, 2e] have described the same essential features.

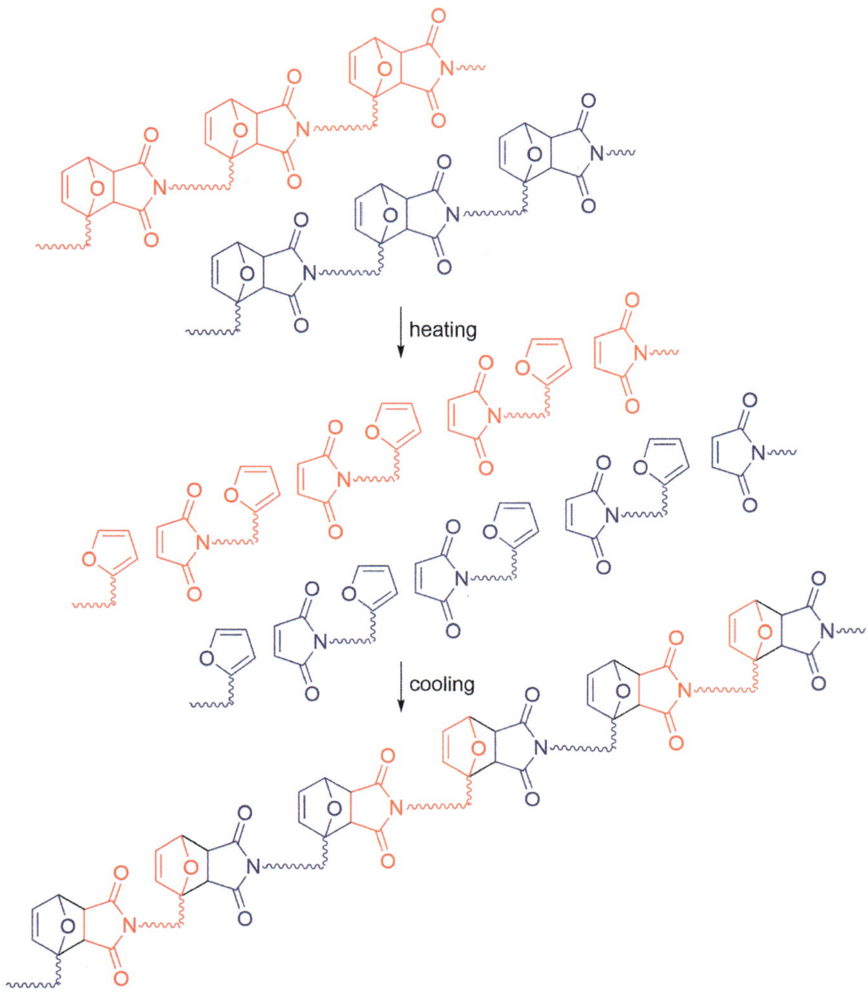

Figure 6.9 Schematic representation of the DA cycle turning the polyester mixture into a copolyester.

Self-healing shape-memory polyurethane networks [12] were synthesized using aliphatic triol and diisocyanate in conjunction with a diol incorporating a furan/maleimide adduct, like those shown in Figure 6.12.

The relevant feature of these materials is that thanks to their shape-memory property, the healing process of a produced crack, induced by the thermal rDA/DA cycle, took place without the need of an external pressure.

DA dendrimers based on the furan/maleimide couple were first reported in 2001 [1, 13] and the study amplified in scope a few years later [1, 14]. In the

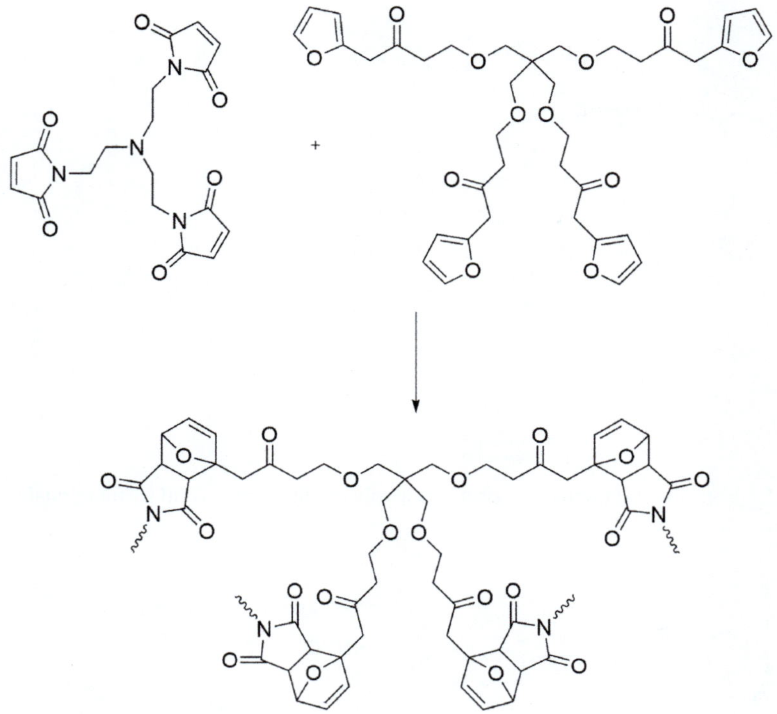

Figure 6.10 DA cross-linking polycondensation between two multifunctional monomers.

original contribution [13], a central aromatic core diverged through three DA adducts bearing two hydroxyl functions, which were etherified with different aromatic branching moieties. Figure 6.13 depicts the structure of the third-generation dendrimer. Other similar approaches were further reported by the same group [1, 14] and other authors, including high-Tg materials [1]. In all instances, the rDA reaction, carried out at 110 °C, was shown to operate smoothly by regenerating the starting dendrons and, likewise the dendrimers were reconstructed, albeit slowly, at 60 °C.

An interesting work called again upon the thermally reversible furan/maleimide DA reaction for the synthesis of dendrimer architectures together with thiol-maleimide *click* reaction [15], but the final structures only bore maleimide terminals. Figure 6.14 shows the principle of this strategy with furan-protected maleimide adducts and the G4 dendrimer after deprotection, and reaction of the SH moiety with maleimide.

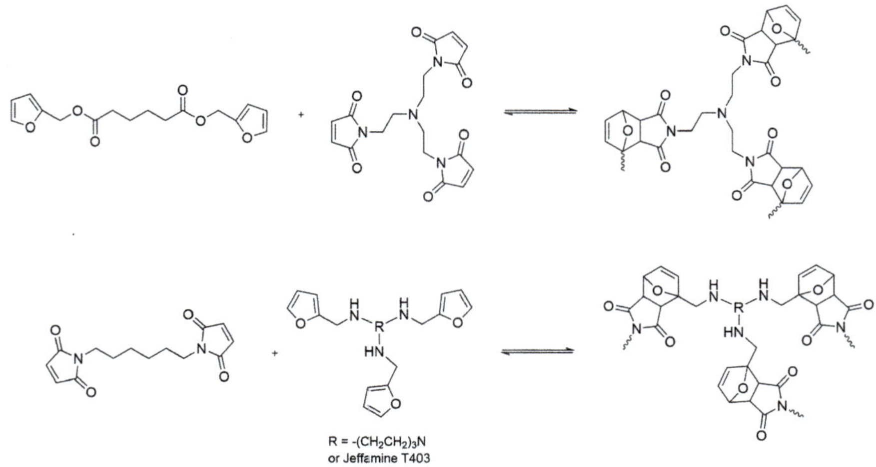

Figure 6.11 Non-linear DA polycondensations involving different di- and tri-functional monomers.

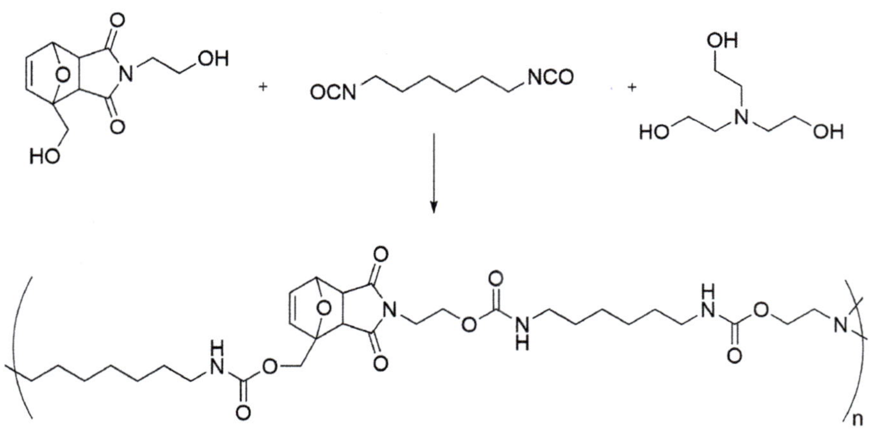

Figure 6.12 The synthesis of a polyurethane network incorporating a furan/maleimide DA adduct.

6.2 Polymer Modification and Cross-linking

A preliminary assessment of the reactivity of furan heterocycles as a function of their incorporation into a polymer structure was carried out using a large selection of the polyurethanes discussed in Section 5.3, some of which are shown in Figure 5.42, since they bore the ring in a variety of situations, including as C2- substituted pendant moieties and as integral elements of the polymer backbone [16]. These DA reactions involved both monomaleimides, which were followed by NMR spectroscopy, and bismaleimides, which were also monitored by viscometry until the

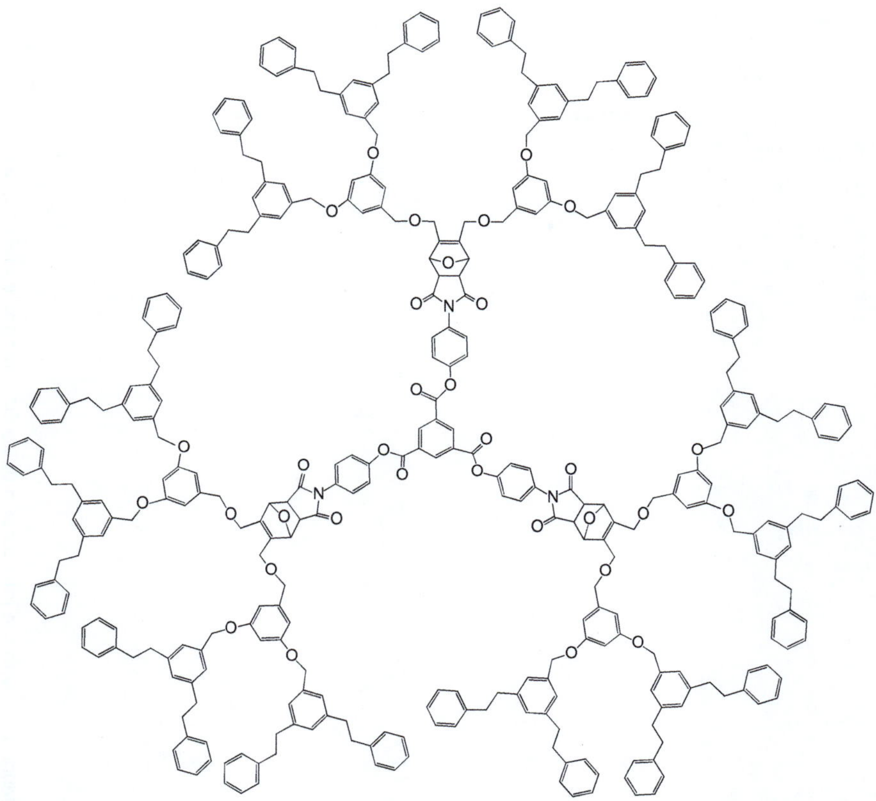

Figure 6.13 The structure of a third-generation furan/maleimide DA dendrimer.

onset of gelling. All DA couplings proceeded following the classical mechanism, but their rates at 40 °C varied significantly depending on the steric availability of the furan ring in the polymers, as qualitatively displayed in Figure 6.15. The same trend was observed when the furan-containing polymers were acrylic macromolecules. These results appear entirely reasonable, given that the growing "sequestration" of the ring within the polymer chain inevitably reduces its kinetic response to the DA coupling with the maleimide reagent, in both simple coupling and interpolymer cross-linking.

The first thorough report of a thermally reversible cross-linking arising from the furan/maleimide DA reaction actually involved two polymers (rather the later most common approach implicating a polymer and a DA bifunctional complementary reagent discussed below), namely poly(oxazoline)s bearing, respectively, furan and maleimide pendant functions [17], as shown in Figure 6.16. The authors were also the first to study the thermally induced decoupling of network, and proved the total recovery of the original linear polymers.

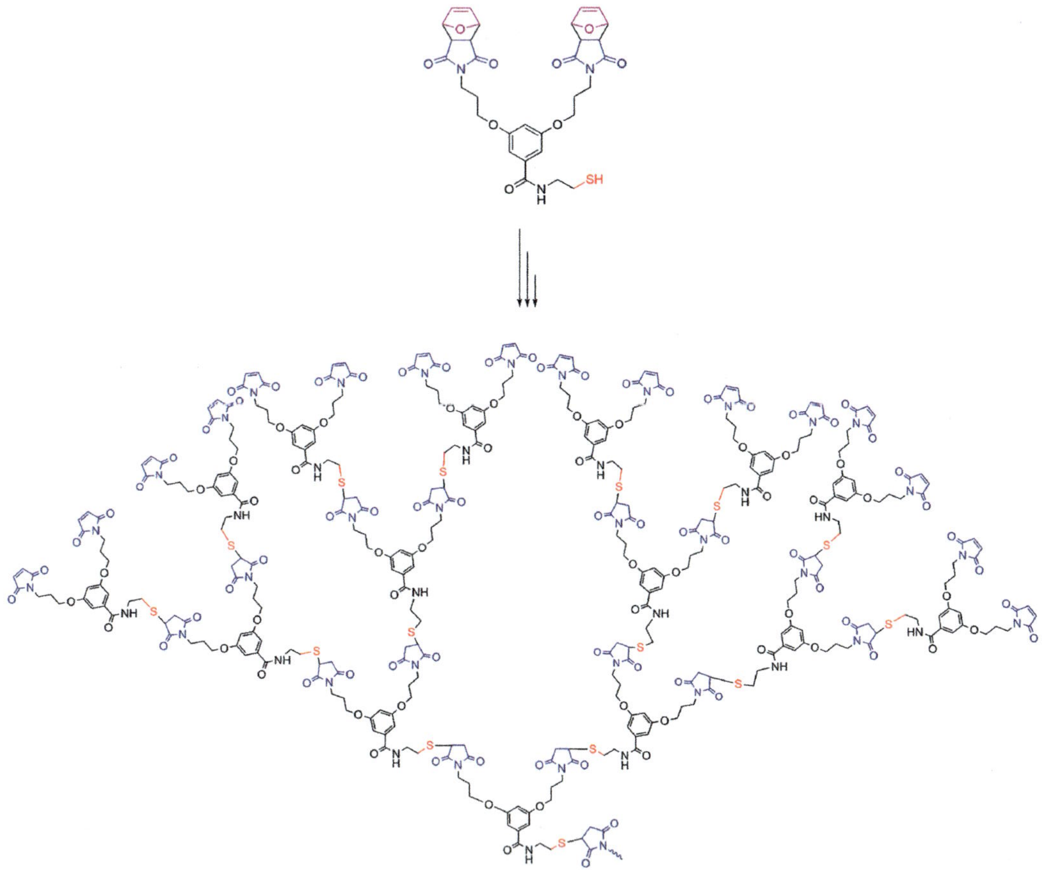

Figure 6.14 Double-click approach to the construction of dendrimers with the intervention of the DA furan/maleimide adduct.

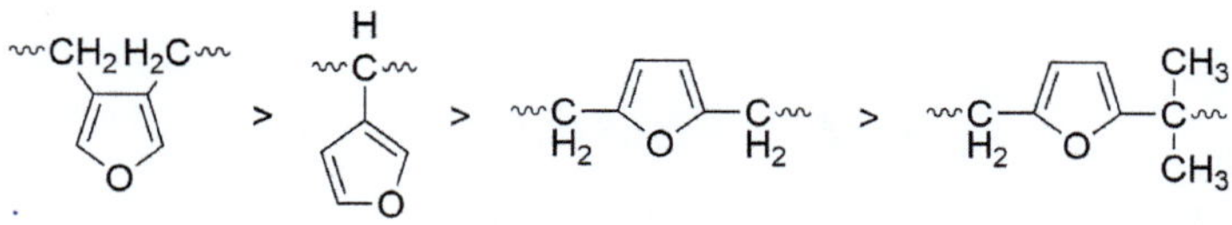

Figure 6.15 Relative DA reactivity of furan moieties borne by different polymers towards mono- and bis-maleimides.

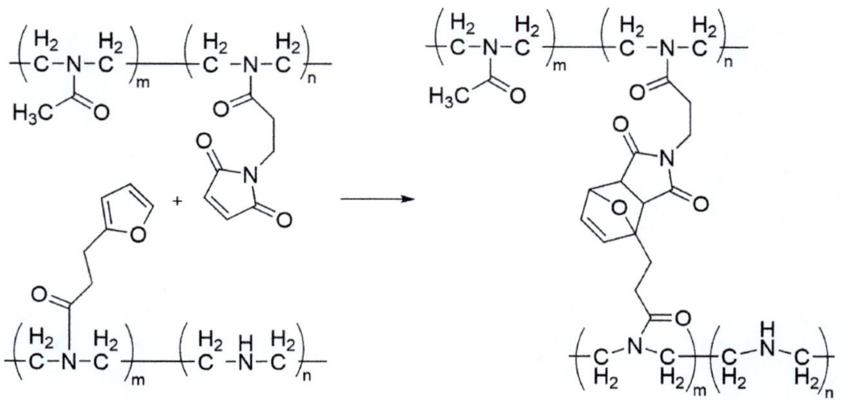

Figure 6.16 DA inter-polymer cross-linking.

A decade later, the cross-linking strategy involving a polymer bearing pendant DA reactive groups and a complementary DA bifunctional counterpart was the object of a double study in which the various facets of these systems were thoroughly investigated [18]. The first system [18a] called upon styrene copolymers with different proportions of a pendant furan comonomer, which were made to react alternatively with an aromatic maleimide and its bis-homologue, as sketched in Figure 6.17. Whereas the former DA condensation yielded the corresponding copolymer bearing the pendant adduct, which was fully characterized, the latter produced the expected gelled material. Both products were then heated at 130 °C in the presence of an excess of 2-methylfuran in order to promote the retro-DA reaction, but avoid that, upon cooling the systems, DA recombination would take place, since the released maleimides had been DA-trapped by the added furan compound.

These experiments were conducted under different conditions, particularly in terms of the copolymer composition (which translated into a linear Tg variation from 100 to 0 °C as the proportion of furan comonomer increased from 0 to 100%), and proved successful in showing the clean and total progress of each reaction, indicating moreover that as little as 5% of furan-bearing comonomer was sufficient to lead to their completion, albeit at correspondingly lower rates. The ensuing DA networks were stiff, with Tg values above room temperature.

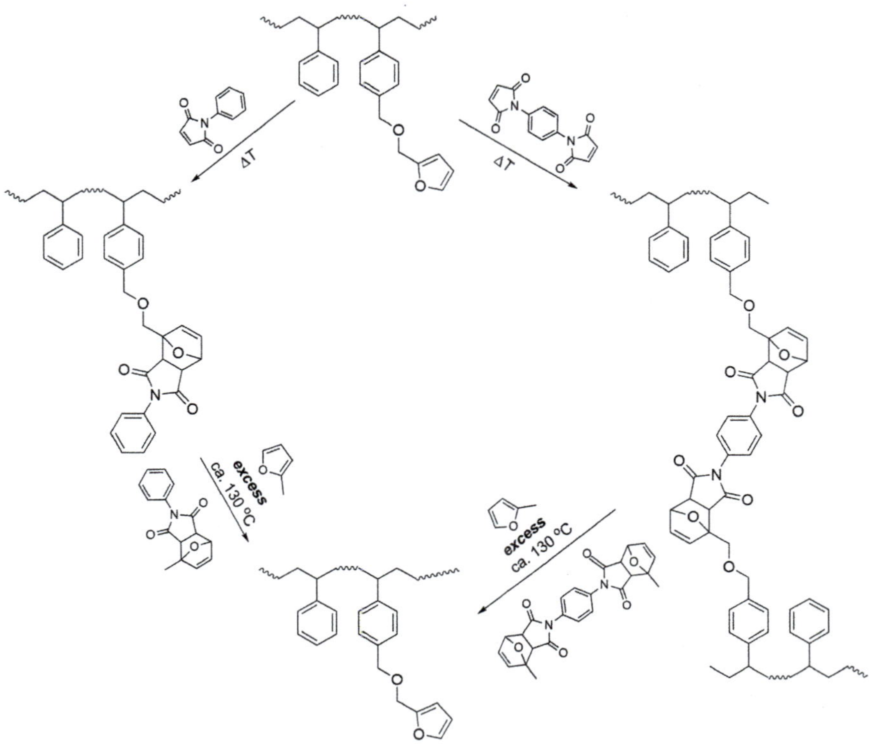

Figure 6.17 The DA forward and reverse reactions involved in mixing a styrene-furan copolymer with mono- and bis-maleimides.

The second system [18b-d] followed the same general strategy, but, on the one hand, used polymers and reagents possessing a softer character in order to produce elastomeric materials, and, on the other hand, two complementary combinations were adopted, viz. dienic copolymers bearing pending furan moieties reacting with bismaleimide dienophiles, and dienophile copolymers with pending maleimide moieties reacting with difuran dienes. The former system is shown in Figure 6.18, where the trap used during the retro-DA de-cross-linking was again 2-methylfuran. Because the acrylic comonomer used here had a long aliphatic side chain, the copolymers were viscous liquids with Tg values decreasing from −29 to −78 °C as its proportion increased from 70 to 97%. The ensuing DA networks were soft elastomers thanks to the high flexibility of the oligoether bismaleimide used to produce the DA cross-linking, and their Tg were only 10 to 15 °C higher than those of corresponding linear copolymers. Again, both the DA cross-linking and its reverse reaction followed the expected behavior, without detectable perturbing reactions [18b-d], confirming that the use of the 2-methylfuran trap during the DA de-cross-linking process insured the clean recovery of the starting linear

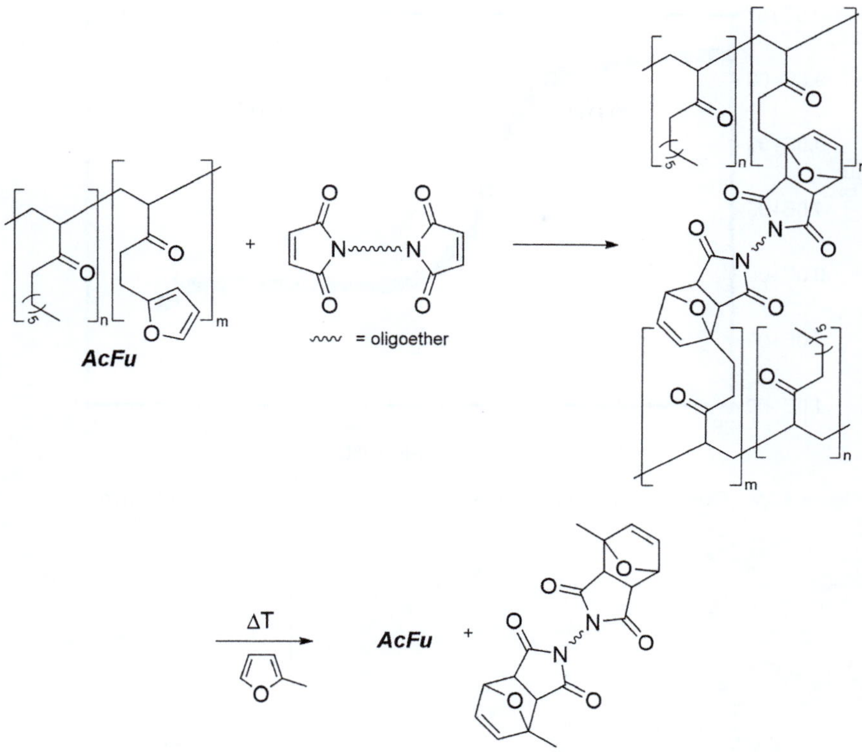

Figure 6.18 DA cross-linking and de-cross-linking of a furan-acrylic copolymer reacting with a bismaleimide.

copolymers. The cycles were successfully repeated to prove the reliability of the chosen strategy. The dynamic mechanical properties of one of these networks [18c] are shown in Figure 6.19, where one can visualize the classical drop in modulus around Tg and, more interestingly, the formation of a viscous liquid at about 100 °C, associated with the onset of the retro-DA reaction, which regenerated the linear furan copolymer and induced the stoppage of the measuring device.

The system used for the complementary approach [18b,d] is illustrated in Figure 6.20. The viscous copolymers with pending dienophile maleimide groups were suitably modified poly(dimethylsiloxane)s and exhibited Tg values of around −110 °C, and the difuran diene cross-linker was also a liquid siloxane reagent. These DA combinations produced soft elastomeric networks with Tg values close to those of the initial copolymers. In this case, the DA de-cross-linking trap was *N*-ethylmaleimide.

The various steps making up this new system were monitored using the same spectroscopic and material characterization as the previous ones [18] and their outcome was found equally satisfactory in terms clear-cut DA responses.

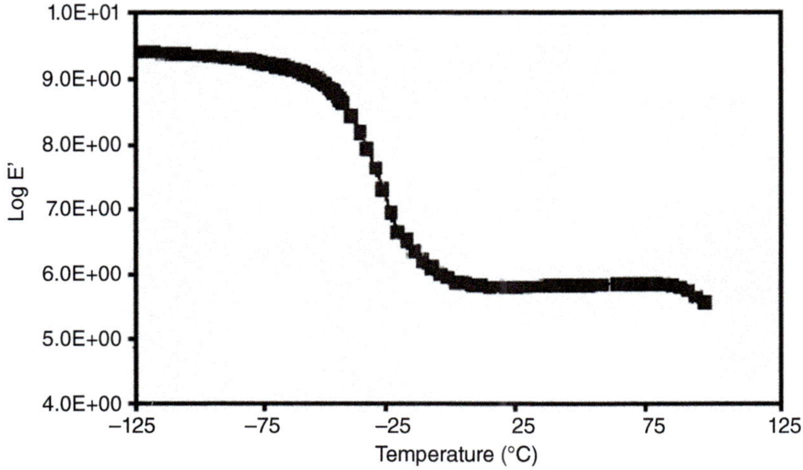

Figure 6.19 Dynamic-mechanical behavior of a DA-cross-linked furan copolymer.

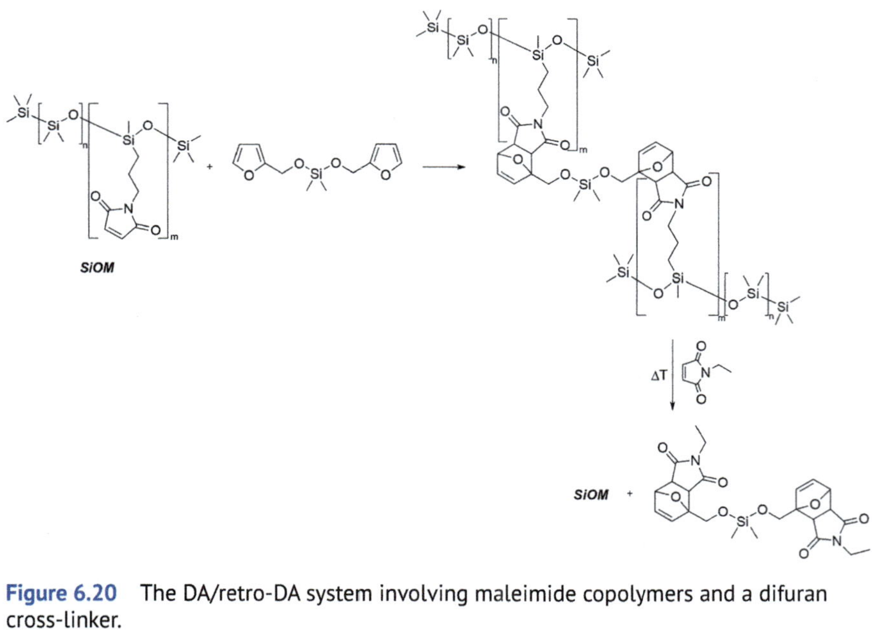

Figure 6.20 The DA/retro-DA system involving maleimide copolymers and a difuran cross-linker.

Following these thorough investigations, the reports on similar cyclic DA processes involving thermally reversible polymer cross-linkings have become progressively more numerous, with the important difference that, as time goes on, they have tended to become more repetitious, lacking the qualitative and quantitative originality of some of the studies reported in the 2013 review [1], in which kinetic features and materials characterizations stood out for their sound contributions to the knowledge of this important topic. In other words, the majority of the more recent investigations only widened to the spectrum of polymers submitted to this approach to the point that virtually all common and less-common synthetic macromolecular materials have been explored as candidates in this context. The most relevant aspects of all these contributions are of course related to the general properties associated with the fundamental characters of recyclability and self-healing and self-mending aptitudes that make the thermally reversible intrinsic feature of these DA cycles so original and useful.

One very relevant area must however be singled out and discussed here in more details, namely the DA modification of natural polymers and macromonomers [19], because it represents the fundamental trend of exploiting renewable resources for preparing the next generation of macromolecular materials.

Polysaccharides have been modified through the furan/maleimide Dies–Alder reaction, as reviewed in 2018 [20] with the purpose of developing novel materials in the areas of self-healing polymers, tissue engineering and drug delivery.

Cellulose, the most abundant natural polymer on Earth, which has been exploited by man as an indispensable material for millennia, has enjoyed a spectacular revival of interest in its recently-developed three nanoscopic forms [21], which have sparkled extensive research and applications in nanocomposites and other functional materials covering a broad range of fields. In 2017, cellulose nanocrystals were used to generate a tough self-healing hydrogel by their surface modification with furfural to give acetal moieties, which were made to react with a poly(ethylene oxide) bismaleimide through the DA reaction [22]. This interesting study had, to our mind, the drawback of the chemical lability of the acetal functions, which are prone to hydrolysis. More recently, a similar strategy [23] was applied to nanofibrillated cellulose bearing superficial carboxylic groups, generated by TEMPO oxidation, which were made to react with 2-furfuryl amine. The ensuing furan-modified nanofibers were cross-linked with oligoethylene oxide bismaleimide to produce a viscous hydrogel by the DA reaction, as shown in Figure 6.21. Heating at 95 °C deconstructed the network, as portrayed in Figure 6.22.

In another investigation, the surface of cellulose nanocrystals was furan-modified by carbamation in DMSO and thereafter cross-linked with a bismaleimide [24], but we think that this medium is not as suitable as the natural aqueous environment used in the previous studies. In a different vein, maleimide moieties were appended onto the surface of cellulose nanocrystals, which were then reacted with furan-modified gelatin to generate a bionanocomposite hydrogel [25].

Lignin is the most abundant natural aromatic polymer on Earth and both its complex chemistry and the numerous applications and present research reports in materials science are topics of high relevance [26]. The few applications of the DA reaction with lignin are very recent and quite interesting. In 2017, soda lignin modified with maleimide moieties interacted with polyfuran reagents in

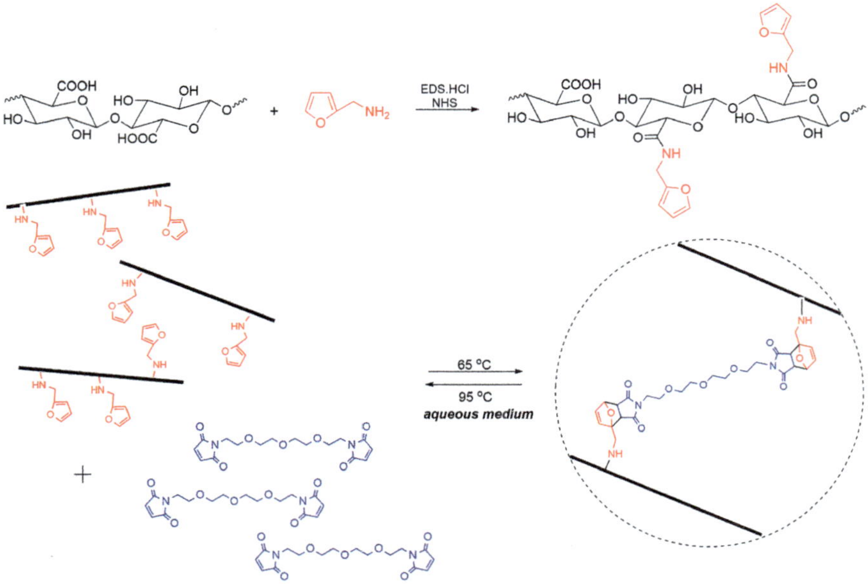

Figure 6.21 Furan-amidation of nanocellulose followed by thermally reversible DA cross-linking with oligoether bismaleimide.

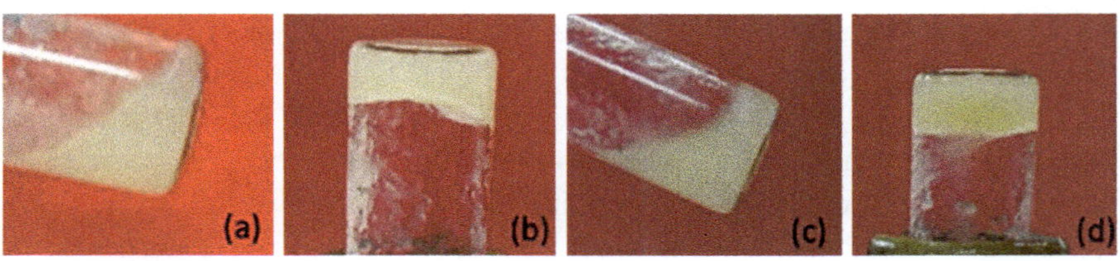

Figure 6.22 The double DA cycle: (a) before cross-linking, (b) after cross-linking (formation of the hydrogel), (c) after de-cross-linking at 95 °C, and (d) regeneration of the gel after cooling.

a green-chemistry mode (no solvent, no catalyst) [27], as sketched in Figure 6.23 where the classical DA cycle is illustrated. Figure 6.24 shows the success of self-healing following one such cycle. The thermos-mechanical properties of these materials could be tuned as a function of the furan reagent used and its stoichiometry with respect to the appended maleimide functions. Among their potential applications, the authors envisaged their use recyclable UV protecting films, given the well-known filter aptitude of lignin structures.

A later similar study [28] employed furan modified Kraft lignin and an aromatic bismaleimide together with a trifuran polyether to prepare a double network in methyl ethyl ketone. The mechanical properties of these materials varied considerably as a function of their composition and phase separation between to

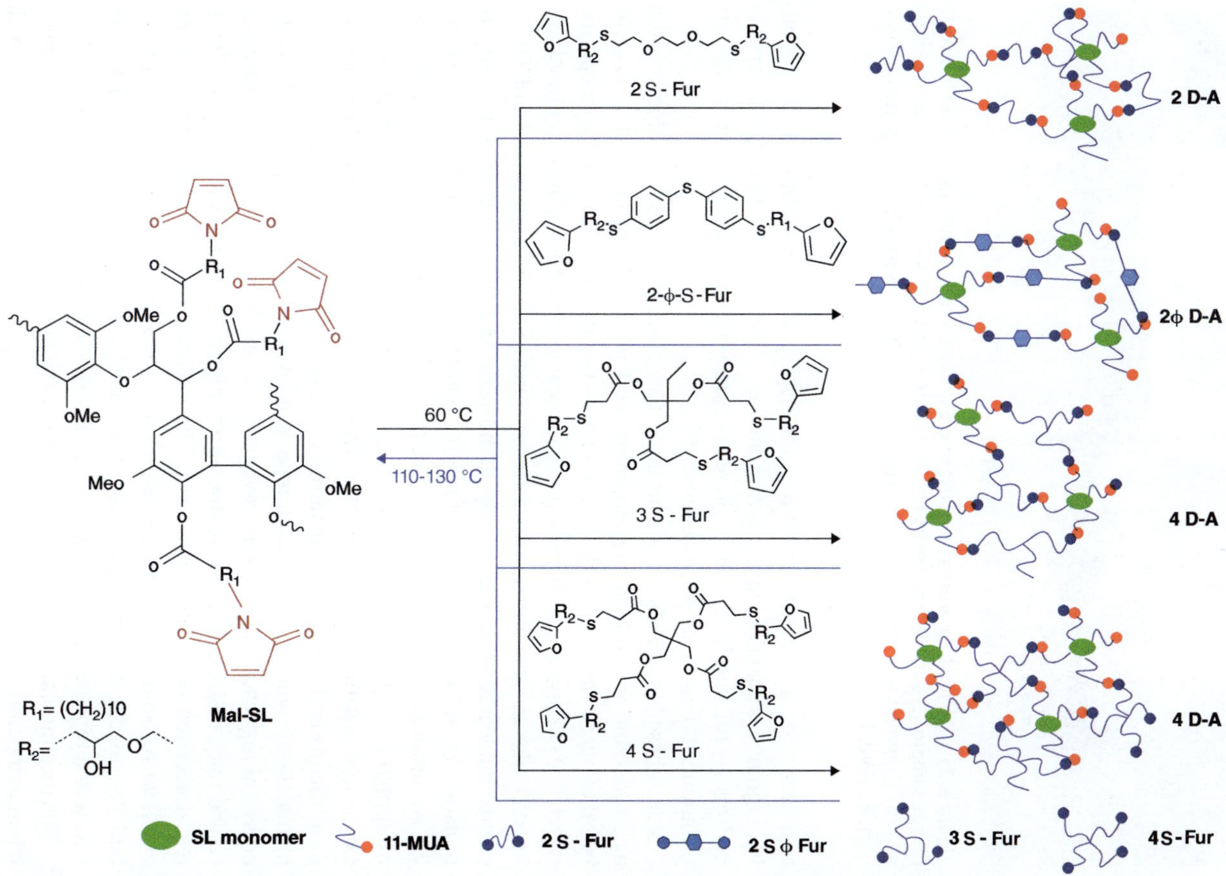

Figure 6.23 Thermally reversible DA cross-linking of soda lignin. Reproduced from Ref. [27] with permission of Elsevier.

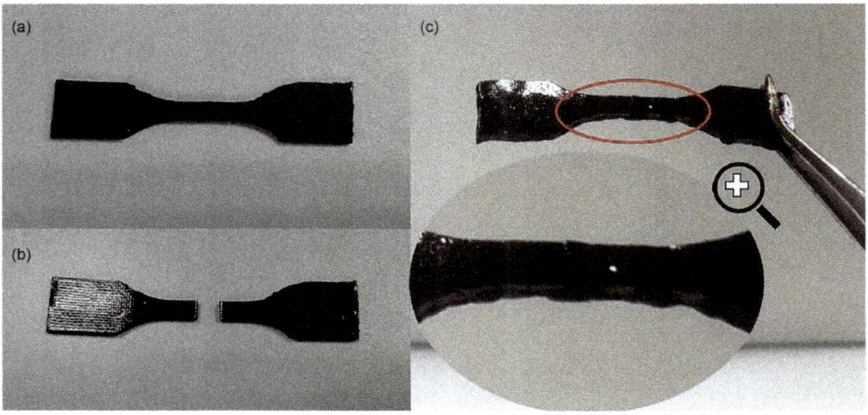

Figure 6.24 DA healing process applied to cross-linked soda lignin: (a) molded network, (b) cut sample, (c) sample fragments healed following retro-DA network de-cross-linking at 120 °C, joining and cooling. *Source:* From Ref. [27], 2021 / Reproduced with permission from ELSEVIER.

two networks, and indeed some did not display a complete thermal reversibility. Lignin was also used as a macromonomer in the elaboration of DA-reversible hydrogels [29], but the impact of these investigations to materials science was less pronounced than those discussed above.

Starch-based materials occupy an important place in renewable polymers, and a considerable research and development work effort is being devoted to optimizing their performances [30]. This includes the development of thermoplastic starch (TPS), copolymers, blends, and chemical modifications, the latter involving in particular the use of reactive extrusion. One of the main drawbacks of TPS, a cheap, biodegradable, and readily manufactured material, is its sensitivity to moisture, which engenders loss of mechanical properties and dimensional stability. An innovative approach to improve this problematic state of affairs consisted in applying the DA cycle to TPS using reactive extrusion in a green-chemistry experimental setup [31]. TPS was first modified by appending furan heterocycles to its chains and the ensuing material was extruded while injecting an aromatic bismaleimide when the medium temperature was above the occurrence of the DA cross-linking reaction, which however took place when the extruded mixture was cooled progressively to room temperature. Figure 6.25 shows the chemical context associated with this experiment. The network thus produced displayed an excellent resistance to water and other polar solvents, compared to the unmodified TPS, as depicted in Figure 6.26. The recyclable character of these materials was tested repeatedly and Figure 6.27 illustrates their aptitude to hot molding.

The mechanical properties of this novel starch network were strongly enhanced when compared with those of its TPS precursor, as shown in Figure 6.28, which also displays the onset of the retro-DA deconstruction at about 140 °C.

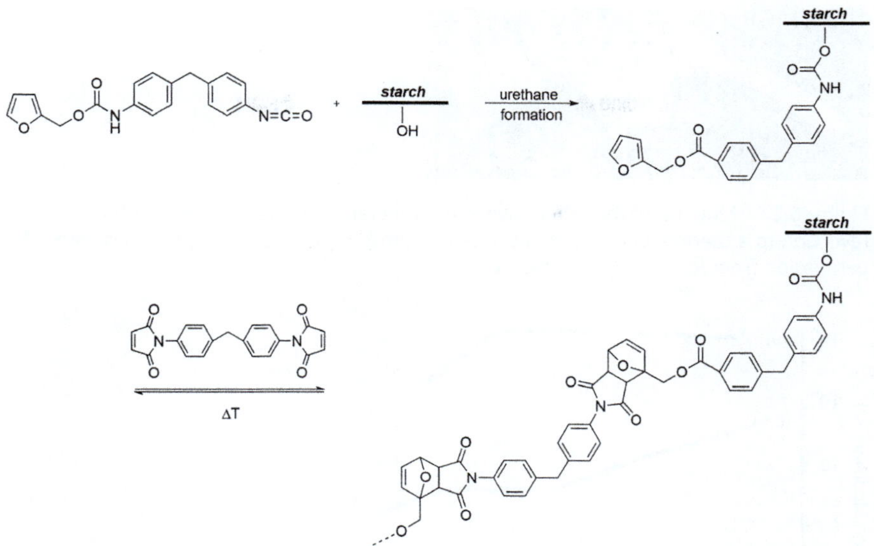

Figure 6.25 TPS furan modification prior to reactive extrusion in the presence of the aromatic bismaleimide.

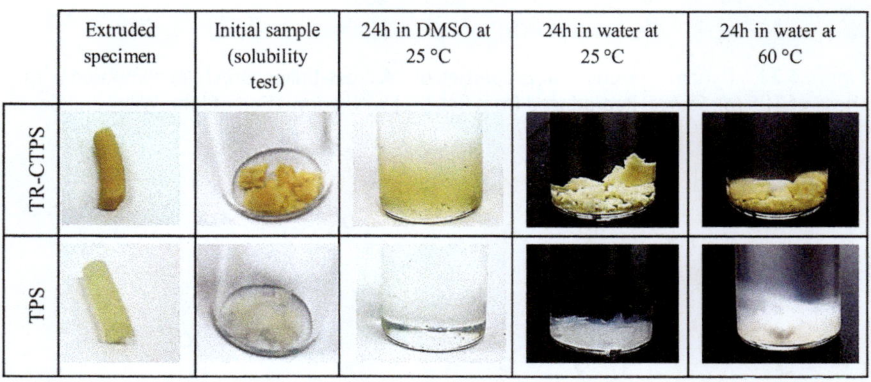

Figure 6.26 Comparison of solvent resistance of conventional TPS versus extruded and DA-cross-linked TPS. *Source:* From Ref. [31], 2015 / Reproduced with permission from John Wiley & Sons, Inc.

In a more recent approach, some of gelled starch's primary OH groups were TEMPO oxidized to carboxylic function before esterification with furfuryl alcohol, and the ensuing material was DA cross-linked with a water-soluble bismaleimide [32], at neutral pH and body temperature, as sketched in Figure 6.29. The main purpose of this study was to develop a cell-compatible gel for regenerative medicine, as successfully assessed by cell cytotoxicity tests. Obviously, there is no reason here to call upon the retro-DA reaction.

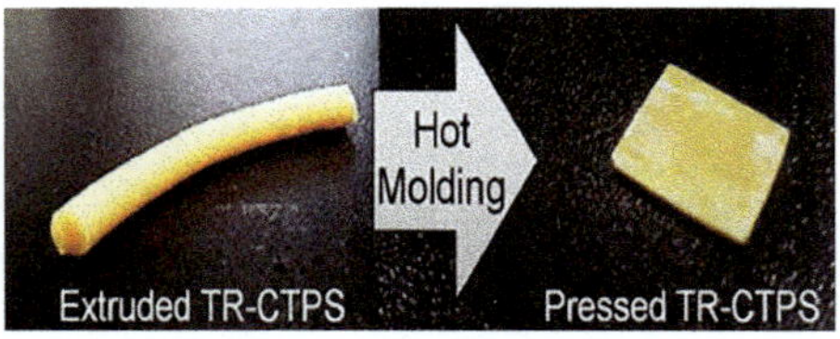

Figure 6.27 Molding of the TPS network at temperature above the onset of its DA reversion to a thermoplastic material. *Source:* From Ref. [31], 2015 / Reproduced with permission from John Wiley & Sons, Inc.

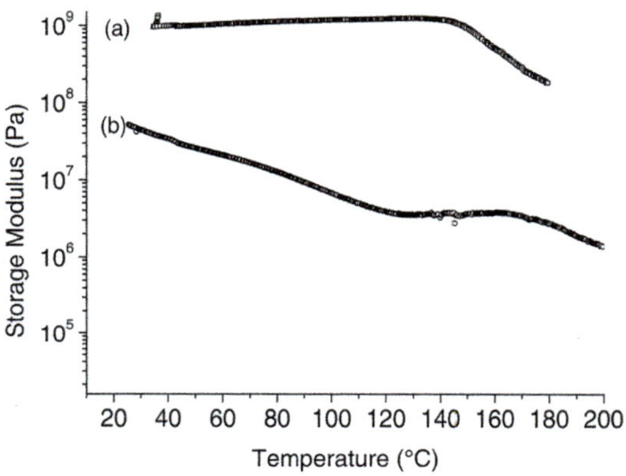

Figure 6.28 Dynamic mechanical properties of DA cross-linked starch (a) compared with those of TPS (b). *Source:* Reproduced from Ref. [31] with permission of John Wiley and Sons.

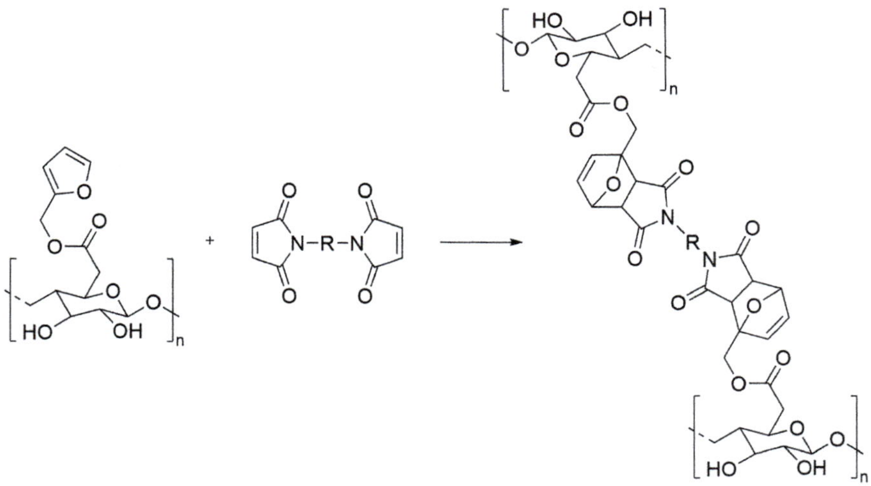

Figure 6.29 DA cross-linking of furan-modified starch with water-soluble bismaleimide in an aqueous medium.

Natural rubber (NR) is regaining in worldwide production after a period of decrease with respect to synthetic counterparts. The major use of NR is destined to tires and conveyor belts, followed by gloves, automotive parts, footwear, adhesives, and medical appliances. The vast majority of these applications require a cross-linking process induced by sulfur, called vulcanization, developed industrially at the end of the nineteenth century, before being also applied, decades later, to synthetic elastomers like poly(butadiene) and its copolymers, and poly(isobutene-co-isoprene), known as "butyl rubber." The basic features of this process are the same today and arguably the most striking of its aspects (both positively and negatively) is the chemical stability of the loose elastomeric networks (typically involving only 3–5 mole % of the monomer units) in terms of recalcitrance to deconstruction, aside from the proneness to material degradation due to atmospheric oxidation of the unsaturated chains. Despite the numerous approaches, no viable industrial process has found the way to recycle tires. This failure has become visible throughout the planet, with mounds of discarded tires, amounting to more than one billion every year. The existing measures to alleviate this huge environmental problem, such as their use in road flooring or combustion for energy recovery, are far too modest to reduce their accumulation in any significant measure. New methods of recycling vulcanized elastomers are therefore urgently needed. The mechanism of the furan/maleimide DA/rDA reaction, with its specific temperature range, appeared as a suitable approach to the solution of the problem of "recycling tires" [33]. The obvious idea was to append furan heterocycles onto the elastomer chains and then cross-link them using a bismaleimide. The thiol-ene click reaction seemed appropriate for the first task, based on the coupling of furfuryl thiol to the unsaturations of the polymer and was first applied to a commercial 1,2-rich poly(butadiene) sample (PBD), typically employed for tire manufacture, since the side vinyl moieties would display a good reactivity in the thiol-ene coupling, as illustrated in Figure 6.30. In order to simulate the degree of cross-linking typically associated with the vulcanization process, the thiol-ene random coupling of furfuryl thiol to the side vinyl group of PBD was limited to about 4% of its monomer units. The modified PBD was then treated with a stoichiometric amount of 1,10-(methylenedi-4,1-phenylene)bismaleimide in DCE at 60 °C and the solution gelled within hours, as indicated by the absence of flow. Upon heating to 110 °C, the medium viscosity decreased progressively, and within hours the network was completely de-cross-linked, as revealed by ^{1}H NMR spectroscopy. This cycle was repeated to confirm the absence of side reactions and to validate the working hypothesis about the possibility of recycling sulfur-cross-linked elastomers [33].

The same concepts were then applied to natural rubber [34] modified with 2% furan pending groups, as shown in Figure 6.31, before successfully cross-linking with the same bismaleimide. Again, the deconstruction of the DA network at 110 °C produced a full return to the reagents within hours, and again the DA/rDA cycle

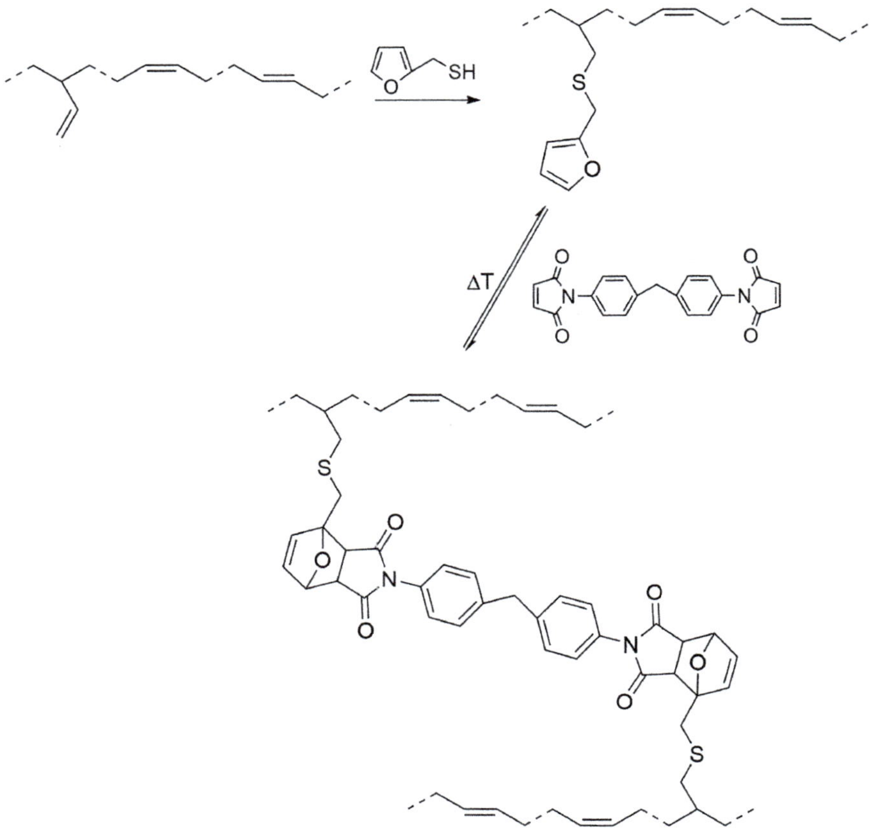

Figure 6.30 DA-reversible cross-linking of poly(butadiene).

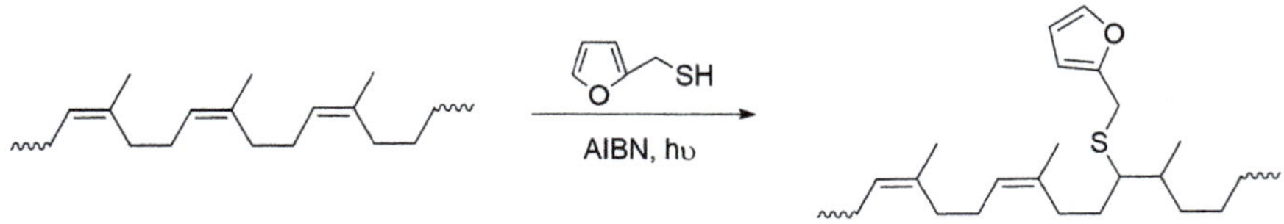

Figure 6.31 Thiol-ene click reaction between natural rubber and furfuryl thiol.

was shown to be reproducible over a few sequences. Figure 6.32 visualizes one such cycle.

This new cross-linking mode enhanced the mechanical properties of the natural rubber (Figure 6.33), as in the case of conventional sulfur vulcanization.

The novel DA-based strategy provided strong evidence for a viable and sustainable solution to the recycling of otherwise chemically intractable conventional sulfur-based network materials, such as tires.

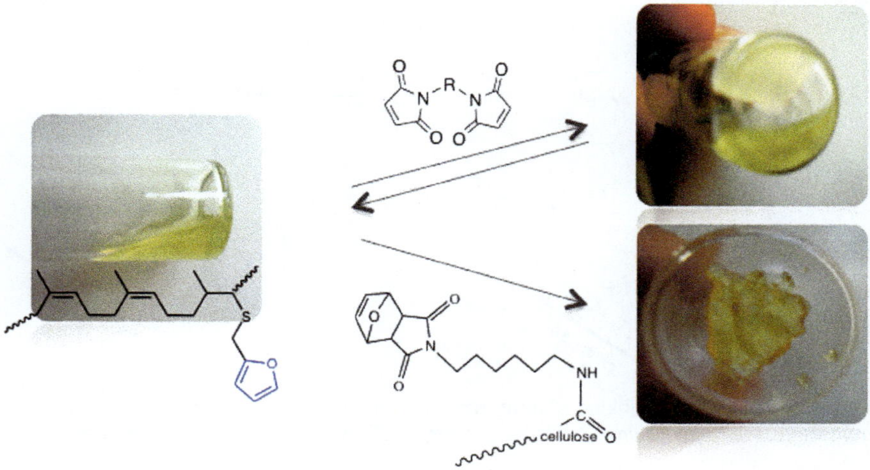

Figure 6.32 Two modes of exploiting the furan-substituted natural rubber. Reproduced from Ref. [34] with permission of Elsevier.

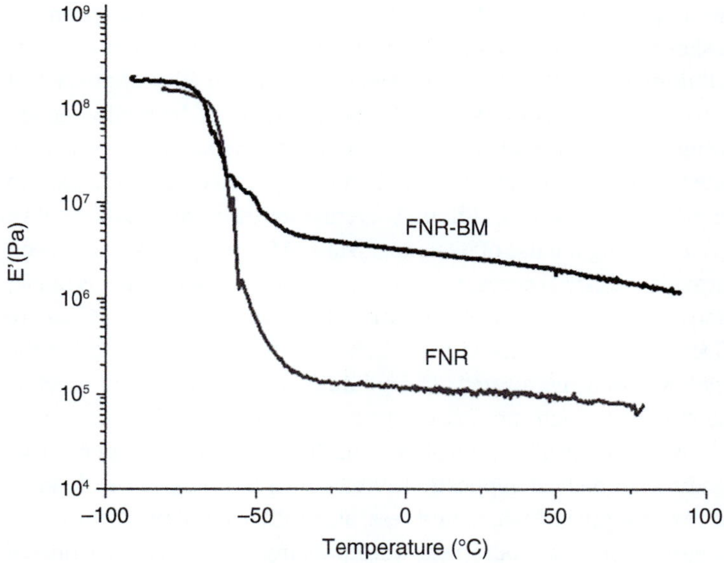

Figure 6.33 Dynamic mechanical properties of DA natural rubber network (FNR-BM) compared with those of its linear furan-modified (FNR) precursor.

In the same study [34], the furan-modified NR was made to react with maleimide-grafted TEMPO-oxidized carboxylic nanocellulose fibers (Figure 6.34) in order to generate a nanocomposite entirely composed of natural polymers. This *chemically bound* composite displayed highly enhanced mechanical properties

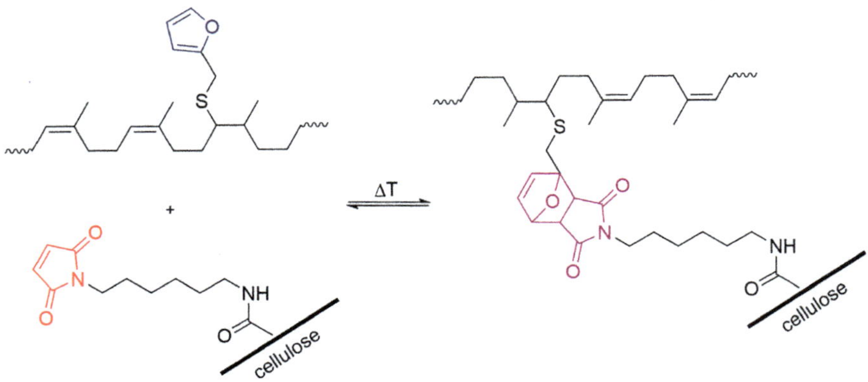

Figure 6.34 DA cross-linking of furan-modified NR with maleimide-modified nanocellulose fibers. *Source:* Reproduced from Ref. [34] with permission of Elsevier.

compared with those of a *physical mixture* of the same modified NR and TEMPO-oxidized carboxylic nanocellulose, as shown in Figure 6.35.

Chitin is the most widespread animal-based polysaccharide and the second most important natural polymer in the world [35]. It constitutes the basic component of the exoskeleton of insects, mollusks, and crustaceans. Its structure resembles that of cellulose, but with a major difference in that the C2 secondary OH groups in the glucose monomer units of the former are replaced by methyl-amide moieties, resulting in an even higher cohesive energy to the polymer because of the stronger interchain C=O···H-N hydrogen bonding. Chitin is therefore hard to dissolve and process into materials, although several interesting applications have been developed, including the use of its nanocrystals [35]. Its hydrolysis has been optimized to provide industrial access to its amino form, chitosan, which is in fact a random copolymer of chitin and amino unis, as shown in Figure 6.36, where "chitosan" reflects all compositions with $X < 0.5$, i.e., when more than 50% of the amide moieties have been hydrolyzed. The solubility of chitosan in mild acidic aqueous media makes it much more apt to transformation into films and other objects, as well as to chemical modification, and hence the large variety of its materials, including biomedical applications which take advantage of its biodegradability, non-toxicity, physiological inertness, and antibacterial properties.

Chitosan-based hydrogel microdroplets were prepared by the DA reaction of furan-modified chitosan (FC) with a water-soluble oligoether bismaleimide (BM), as sketched in Figure 6.37 [36]. The droplets were highly viscoelastic and displayed a tendency to develop micropores for water outlet and inlet, which could therefore be exploited as delivery vehicles, in the present context for drug release.

Gelatin is another natural polymer which has been submitted to hydrogel formation through the DA reaction [37]. Furan rings were appended to gelatin, and the ensuing product DA cross-linked with the same oligoether bismaleimide in an aqueous medium to generate hydrogels of different compositions. The occurrence

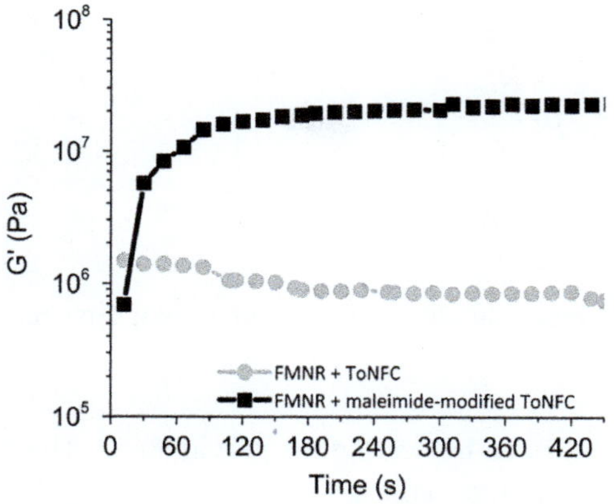

Figure 6.35 Comparative effect of mixing time and storage modulus values of furan-modified NR (FMNR) being cross-linked with maleimide-modified nanocellulose and FMNR physically mixed with oxidized nanocellulose. *Source:* Reproduced from Ref. [34] with permission of Elsevier.

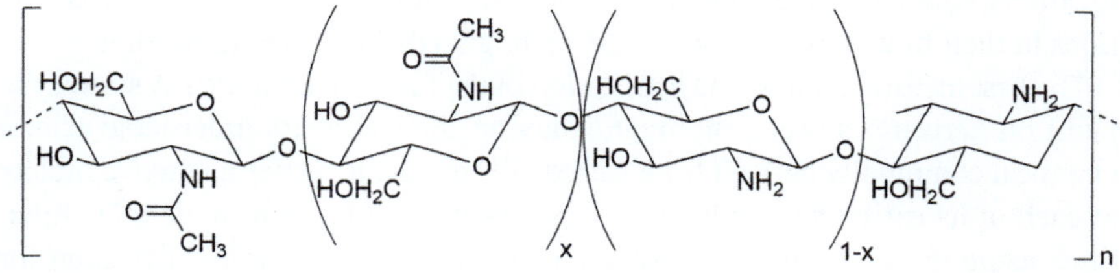

Figure 6.36 The generic structure of chitosan.

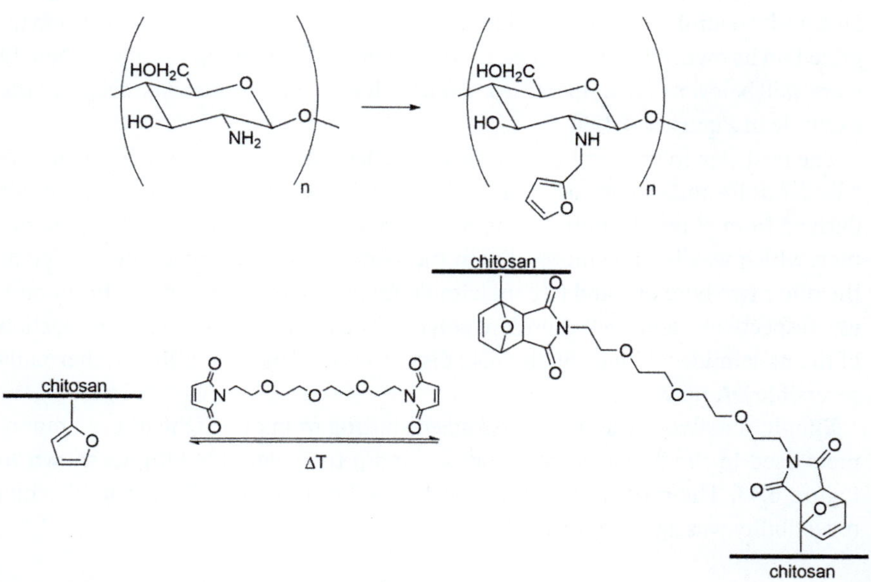

Figure 6.37 DA cross-linking furan-modified chitosan with oligoether bismaleimide.

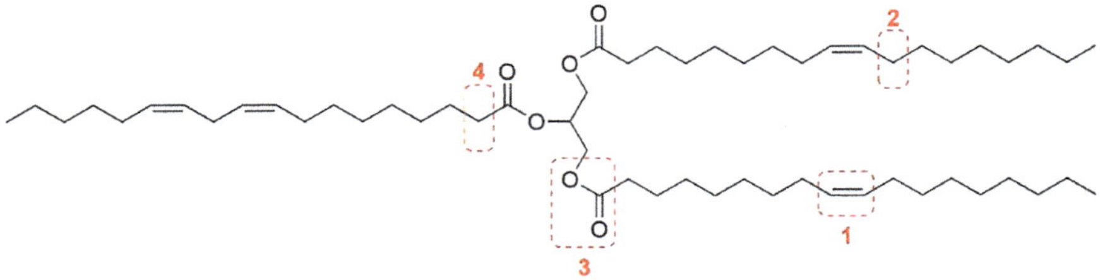

Figure 6.38 An example of triglyceride molecule with the four moieties most exploited for chemical modifications.

of the retro-DA process was also confirmed by heating the gels at 90 °C. These materials are compatible with biomedical applications.

Plant oils, used as such or after specific chemical modifications, and their polymers, are today a rich source of research within the realm of macromolecular materials derived from renewable resources [38]. Among the considerable variety of structures of these triglycerides, those which are best adapted to chemical modifications, with particular emphasis to polymer synthesis, contain C = C unsaturations in their branches and Figure 6.38 highlights the typical reactive sites.

The first investigation on the application of the furan/maleimide DA strategy to plant oil derivatives was a thorough study on the use of 10-undecenoic acid, a chemical commodity derived from castor oil, because it carries a reactive moiety at each of its extremities, suitable for the insertion of furan or maleimide functions using thiol-ene click chemistry for the unsaturation and esterification for the carboxylic group [39]. Three difunctional monomers were thus synthesized, as shown in Figure 6.39, two bearing furan end-groups, which were DA polymerized with an aliphatic bismaleimide, and one AB structure, which was DA polymerized on its own, all giving linear, thermally reversible macromolecules. Their Tg were well below room temperature, given their long aliphatic backbones, as in the example in Figure 6.40 [39].

The next step in this strategy was to synthesize trifunctional monomers, always based on 10-undecenoic acid, in order to prepare DA cross-linked materials derived from plant oils [40]. Figure 6.41 shows the synthesis of a trifuran monomer, which was DA polymerized with the same aliphatic bismaleimide, whereas the other two bore one and two maleimide functions and two and one furan moiety, respectively, and were therefore polymerized on their own, after deprotection of the maleimide adducts. Figure 6.42 displays one of these non-linear, thermally reversible DA networks.

Finally, a network was also constructed starting from one of the difuran monomers used in the linear polymerizations and a trismaleimide [40], as shown in Figure 6.43. The cross-linked material had a Tg of about 0 °C and its thermal reversibility was again confirmed.

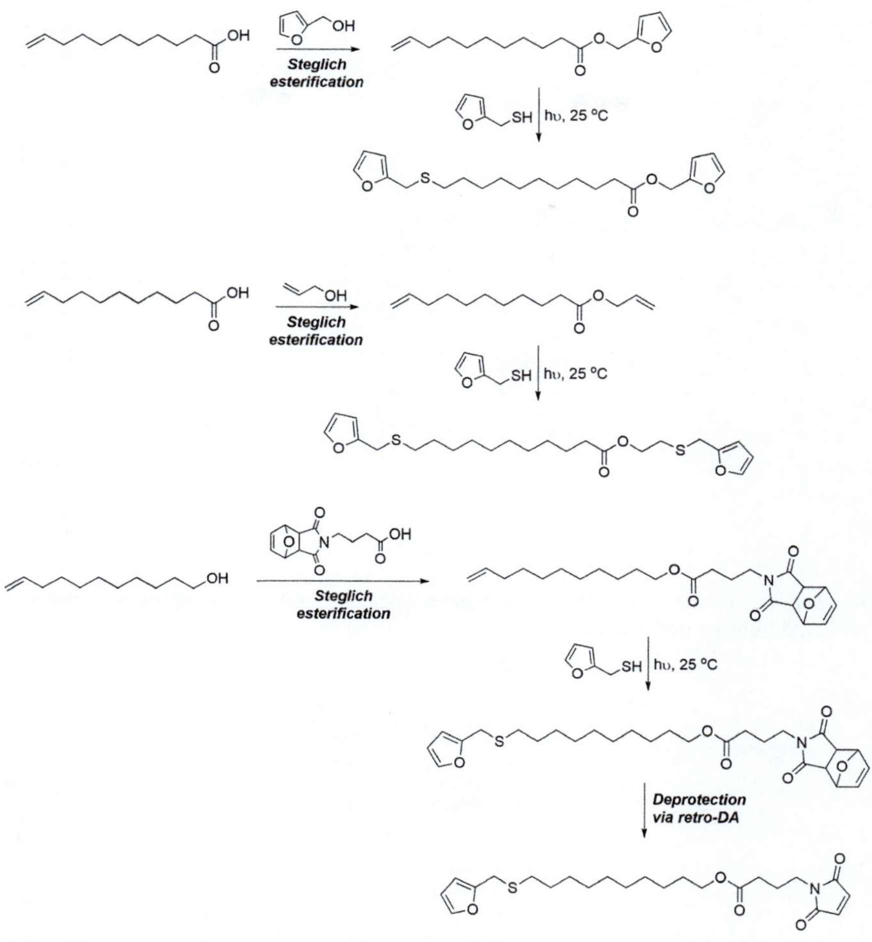

Figure 6.39 Synthesis of three monomers derived from 10-undecenoic acid suitable for DA polycondensation.

The use of triglycerides themselves was also pursued by calling upon a commercial epoxidized linseed oil containing an average of five oxirane moieties per molecule, which was treated in bulk with an excess of 2-furfuryl amine [41], giving rise to two concomitant reactions, viz. the transamination of the ester groups and the oxirane ring opening, as shown in Figure 6.44. The ensuing fragments contained either two or three furan heterocycles and therefore constituted interesting monomers for DA polycondensations with an aromatic bismaleimide. These systems yielded a mixture of linear and branched polymers with Mw values of some 40 kDa and Tg temperatures of 80–100 °C without the formation of networks because of the predominance of difuran monomers over the tri-substituted

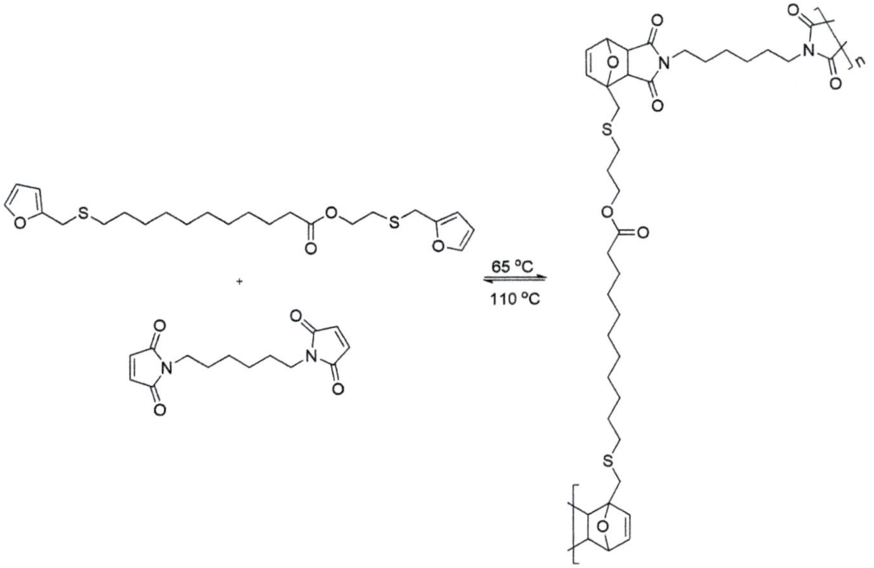

Figure 6.40 Linear DA thermally reversible polycondensation of one of the monomers prepared from 10-undecenoic acid.

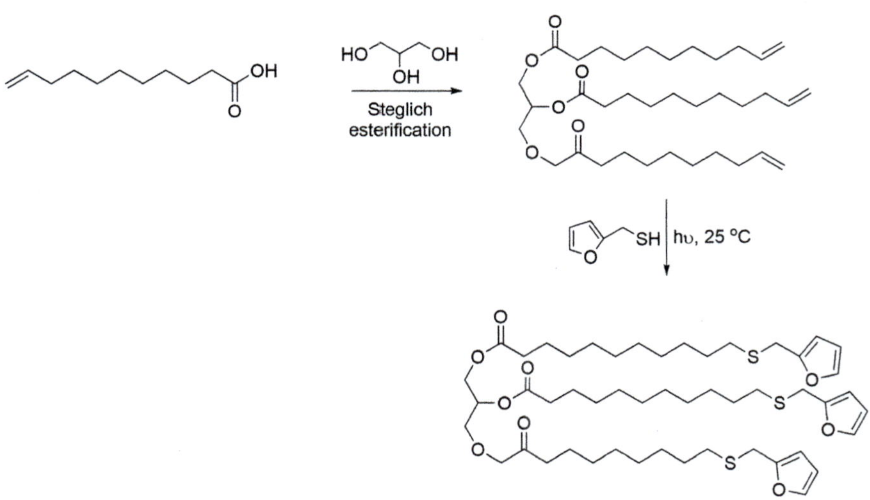

Figure 6.41 Synthesis of a trifuran monomer from 10-undecenoic acid.

homologues. The retro-DA reaction produced in all instances the original monomer mixture, which were readily repolymerized at 65 °C. The application of the same procedure to commercial epoxidized soybean oil gave rise to the same overall behavior and results [41].

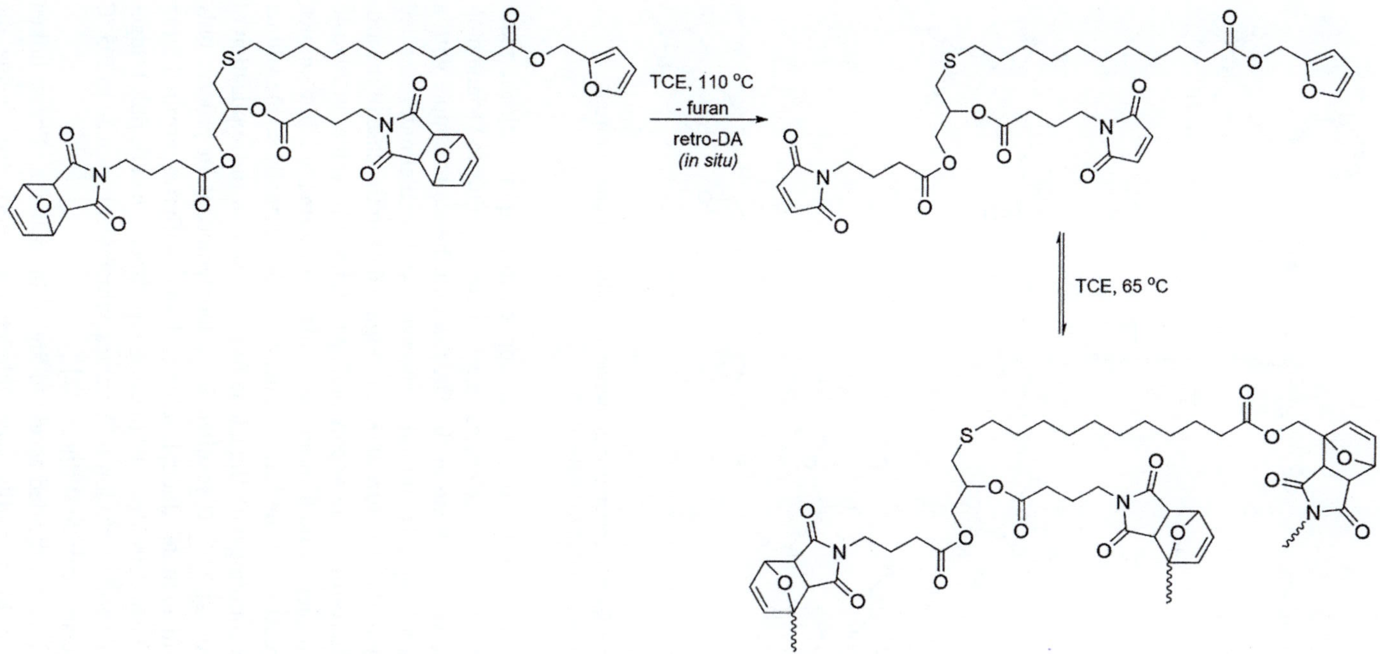

Figure 6.42 Non-linear DA polymerization of an AB_2 monomer based on 10-undecenoic acid.

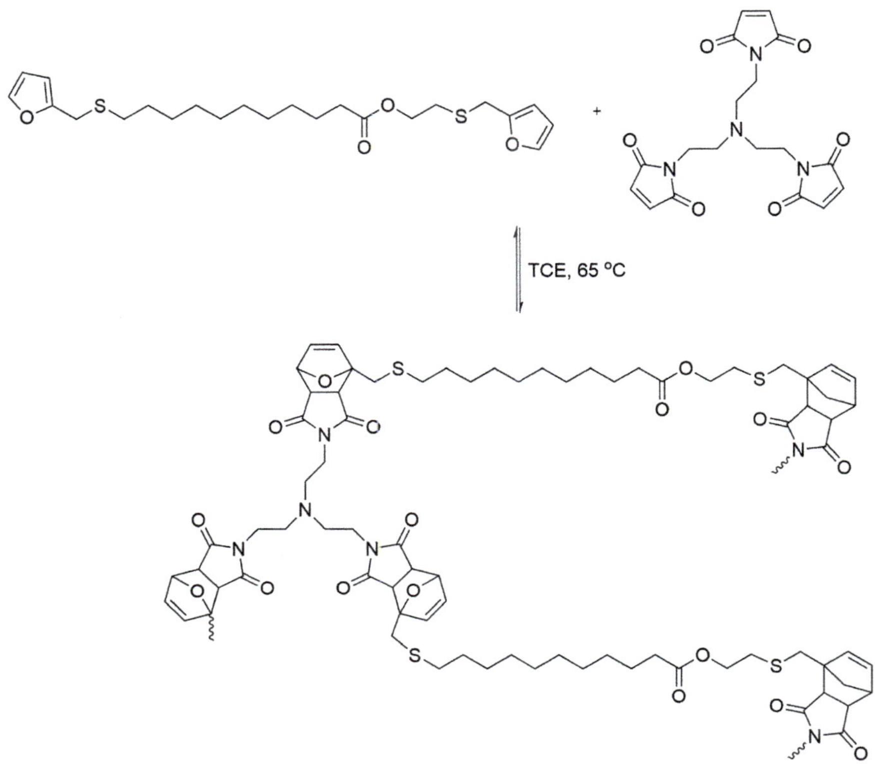

Figure 6.43 Non-linear DA polycondensation between a difuran monomer from 10-undecenoic acid and a compact trismaleimide.

When the same amidation treatment was applied to tung oil [42], the main product was the furan amide of α-eleostearic acid, whose conjugated triene moiety is known to behave as a diene in the DA reaction, hence to couple with a maleimide function. Mixing this tung-oil derivative with bismaleimides resulted in linear polymers with the structure shown in Figure 6.45 in the case of the aromatic dienophile. Interestingly, these polymers, which had Mw values of 20 to 40 kDa, incorporated in each of their monomer units one thermally labile adduct, formed by the DA reaction of the furan diene, and one thermally stable adduct arising from the DA coupling with the triene function, so that the application of the retro-DA reaction at 110 °C resulted in a depolymerization process only involving the furan/maleimide adduct leading to dimers with two terminal furan rings. The other two bismaleimides used in these experiments had flexible bridges and therefore gave networks with lower Tg values, which went from 30 to 105 °C as a function of the bismaleimide stiffness [42].

On the whole this work provided access to linear, branched and cross-linked polymers constructed from readily accessible plant-oil derivatives using the

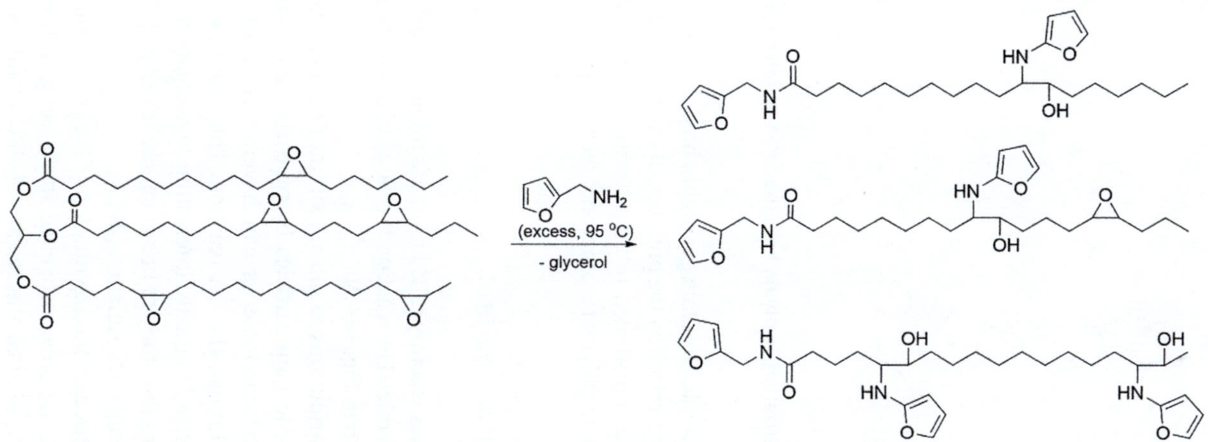

Figure 6.44 Transamination and oxirane opening in the bulk reaction of epoxidized linseed oil with an excess of 2-furfuryl amine.

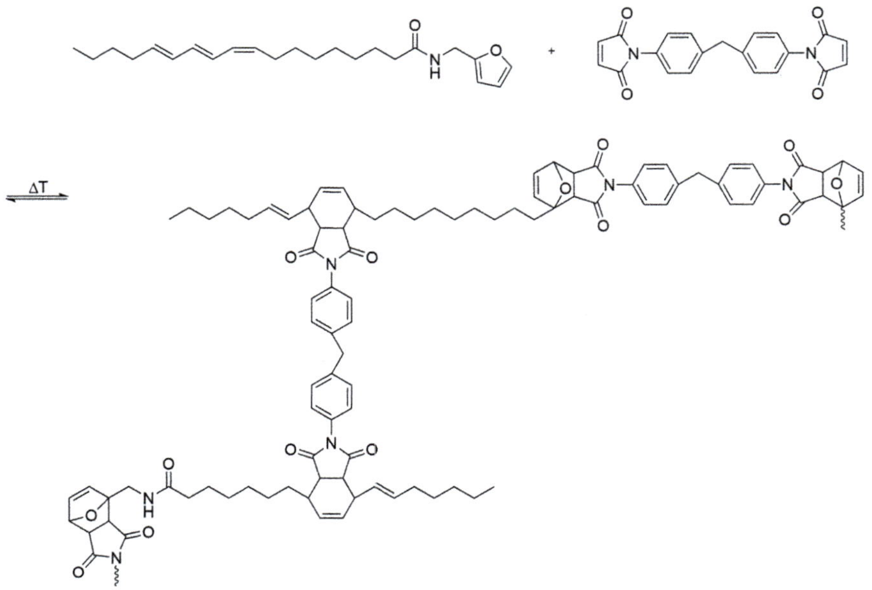

Figure 6.45 DA polymerization involving both the furan and the "triene-type" dienes.

reversible furan/maleimide DA strategy, highlighted the possibility of preparing low-Tg materials characterized by the presence of extended aliphatic sequences in their architecture, and contributed to the increasing exploitation of renewable resources in the realm on polymer science and technology.

6.3 Miscellaneous Systems

One of the areas that has recently called most attention has to do with the application of the furan/maleimide DA reaction is related to biomedical field [43], covering the domains shown in Figure 6.46.

One of the first contributions to this important field [44] showed how to generate immune-polymeric nanoparticles by preparing macromolecular micelles surface-decorated with furan moieties, which were then coupled with maleimide-modified antibodies through the DA reaction. This fine biomedical application was followed by a series of equally interesting investigations in which the DA reaction was applied to the study of breast cancer invasion [45], tunable 3D cell cultures [46], and 3D cell encapsulation [47].

Much has been reported on the exploitation of the DA reaction for drug delivery [43, 48, 49], including the previously cited study using chitosan [36]. One such early investigation [50] studied the release of PEO chains from gold nanorods

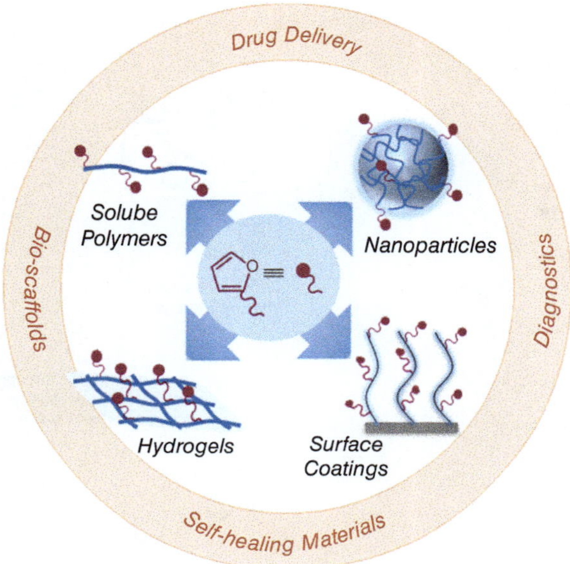

Figure 6.46 The domains covered by biomedical research on the application of the furan/maleimide DA reaction. *Source:* Reproduced from Ref. [43] with permission of Elsevier.

which had been anchored via a DA adduct. Near IR light was then shined on the nanorods to heat them locally and thus induce the retro-DA reaction that released the PEO-maleimide chains. The aim of this study focused on medical applications like the controlled release of drugs. Since then, the field has prospered with studies such as the use of doxorubicin-conjugated mmune-nanoparticles for intracellular anticancer drug delivery [51], the controlled antibody release from degradable thermoresponsive hydrogels [52], and the elaboration of redox-responsive degradable hydrogels for protein release [53], as schematized in Figure 6.47.

Iron oxide nanocubes have been used as a substrate to polymers incorporating DA adducts, which were deconstructed by a magnetically induced hot spot technique and thus triggered drug release [54], as shown in Figure 6.48, where the maleimide carries the drug "R" and the furan is attached to a polymer carrier.

The attachment of polymers incorporating DA adducts to various substrates including glass, silicon oxide [55] and electrospun nanofibers [56] has been exploited to immobilize, for example, bioactive molecules or facilitate cell growth.

A study described the solid-phase synthesis of a furfuryl-substituted poly(styrene) resin [57], which was shown to be quite effective in capturing and thermally releasing maleimide compounds under conditions compatible with intermediate synthetic steps, including peptide coupling techniques.

A clever trick was proposed and successfully implemented [58] aimed at erasing simply and efficiently the roles played by a surfactant in an aqueous medium

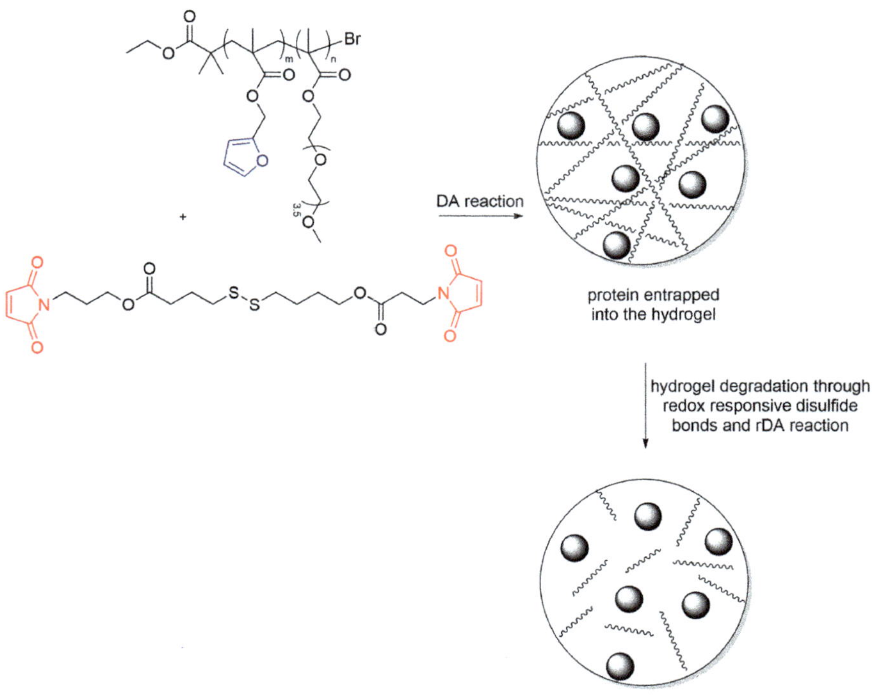

Figure 6.47 The application of the DA reaction for controlled protein release. *Source:* Adapted from Ref. [53].

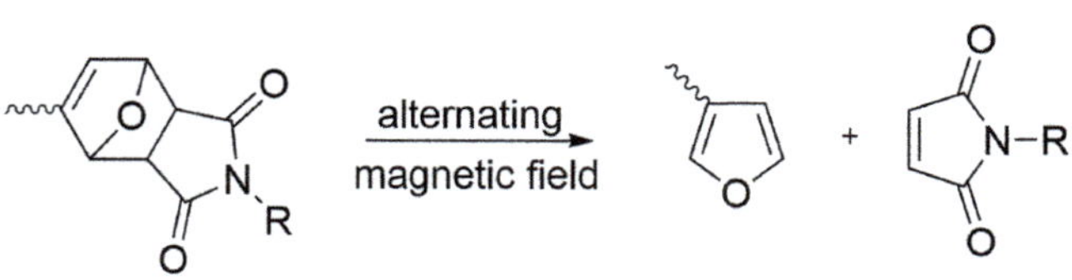

Figure 6.48 Hot spot (magnetic hyperthermia) generated by an alternating magnetic field used to locally decompose the DA adduct and release a drug R.

when these roles are no more required. They synthesized anionic surfactant structures in which the polar and non-polar moieties were joined by a furan/maleimide DA adduct. Their classical features of reducing the surface tension and generating micelles were thoroughly assessed and then the corresponding aqueous systems were heated at 95 °C to initiate the retro-DA reaction of the linker. Within one hour, the surface tension grew from about 55 mN m^{-1} to the water value of 72 mN m^{-1} and, in the case of media containing micelles, the loss of these aggregate structures within 2 hours was detected by SANS measurements. The splitting of the two components of the surfactants by the decoupling of the DA adduct, thus destroying their basic function, was clearly proven by these results. Recombination of the fragments by the DA reaction was made extremely unlikely because of the very low concentrations involved and because the furan-bearing C12 hydrophobic fragment was insoluble in water. The control of surface energy constitutes another promising topic in this context. Polar grafts terminated by furan moieties were attached to the surface of a glass slide using siloxane chemistry and then coupled with perfluorinated compounds bearing maleimide functions through the DA reaction [59]. As a result, the ensuing modified surface displayed a strong hydrophobic character with water contact angles close to 100°, and the subsequent application of the retro-DA reaction by refluxing toluene removed the fluorinated maleimide molecules and regenerated a hydrophilic surface.

In order to improve the performance of polymer-based light emitting diodes, a poly(fluorene) bearing furan pendant groups was prepared and DA cross-linked with two bismaleimides, one of which incorporated a red emitting donor-acceptor-donor bridging moiety [60]. The ensuing devices displayed increased poly(fluorene) emission and enhanced luminance and luminous efficiency. Another notable development showed that furan can be caged by a supramolecular capsule, which renders it unavailable for reacting with maleimide, but can be expelled from the cage by a competing guest like benzene [61].

A systematic investigation of the mesogenic potential of 2,5-substituted furan moieties coupled with 1,4-substituted aromatic counterparts showed that the non-180° alignment of the heterocycles inhibits or strongly limits the possibility of mesophase formations, which are otherwise attained with the aromatic sequence alone [62]. On the other hand, the furan/maleimide DA adduct has been successfully employed to prepare surface-aligned liquid crystal morphologies by the DA linear polymerization of mesogenic monomers incorporating furan and maleimide terminals [63]. These materials have the potential of displaying such properties as actuation, shape memory and surface pattering through a simple synthetic procedure.

Mechanochemistry deals with chemical transformations under the effect of mechanical forces and has witnessed a sustained progress in recent years thanks to its clean methodology and the varied possibilities of original syntheses of both small molecules and macromolecular materials [64]. Here, only its role in replacing heating in breaking the furan-maleimide DA adduct will be briefly discussed. Sonication of molecules and macromolecules incorporating this adduct has been extensively studied, as recently examined in detail, including the different response of the endo- and exo-conformers [65]. Figure 6.49 shows the selective mode of adduct breakage when the mechanical energy applied to this polymer solution does not exceed the threshold that would also induce the macromolecule ester-moieties scission [66].

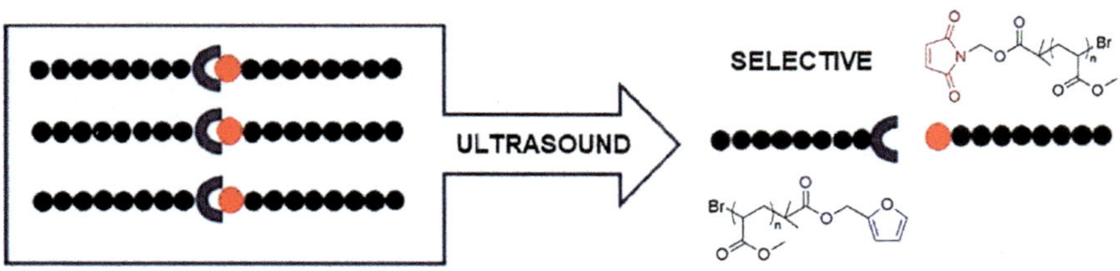

Figure 6.49 Mechanical retro-DA reaction applied to a polymer incorporating a furan/maleimide adduct. *Source:* Reproduced with permission from Ref. [66]. Copyright 2017 American Chemical Society.

In certain systems, the mechanically-triggered adduct reversion to furan and maleimide terminals is accompanied by further polymer chemical disruption, including unzipping back to the corresponding monomer [67]. Conversely, the adduct has been shown to receive mechanical shielding from breaking if connected to a catenane-type structure that can dissipate some of the applied force, but only until full extension of the protecting rings, as shown in Figure 6.50 [68].

Covalent adaptable networks prepared by the DA cross-linking principle and containing a mechanochromic structure have been prepared in order to follow both the internal mechanical damages caused by stress and their random repair by the heating/cooling cycle that rebuilds the DA adducts, as illustrated in Figure 6.51 [69]. As already reported for numerous systems, as previously discussed here, this cycle could be repeated several times.

Ring-opening metathesis polymerization (ROMP) has been applied to different furan-based FA adducts. A series of studies describe the use of modified furan/maleic anhydride adducts as monomers for ROMP using osmium or ruthenium

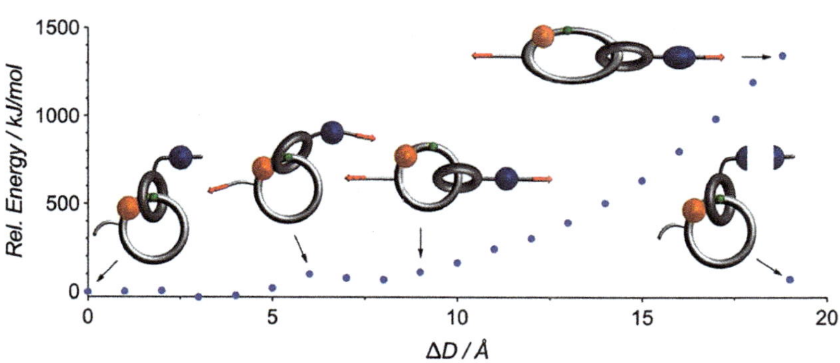

Figure 6.50 Stretching a polymer incorporating a "protecting" catenane joined to a furan/maleimide adduct all the way to the breaking of the latter when the tension on the former reaches a maximum.

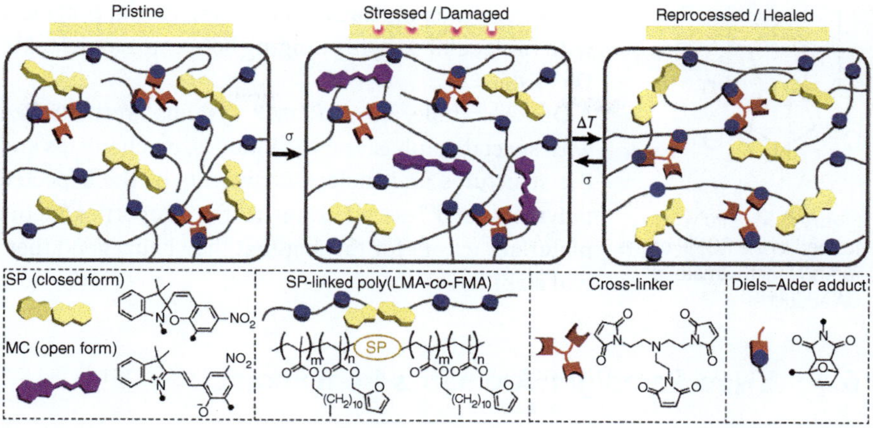

Figure 6.51 Stress-induced DA breaking and temperature cycle healing in a polymer network incorporating furan/maleimide DA adducts. *Source:* Reproduced from Ref. [69], under Creative Commons Attribution 4.0 International License (http://creativecommons.org/licenses/by/4.0).

trichloride as catalysts, and gave different macromolecular structures [70], as indicated in Figure 6.52. After reduction of this adduct, the corresponding polymer is also given in Figure 6.52. The same initial adduct was hydrolyzed and its ROMP [71] gave the polymer shown in Figure 6.53, with Mn values higher than 10^5 Da. Likewise, the ROMP of the DA adduct of furan with *N*-methylmaleimide gave the macromolecular structure displayed in Figure 6.54 with molecular weights ranging between 10^5 to 10^6 Da [72].

The furan/maleimide DA adduct was also submitted to ROMP, but this time in an aqueous-methanol medium, i.e., in a simpler and more sustainable

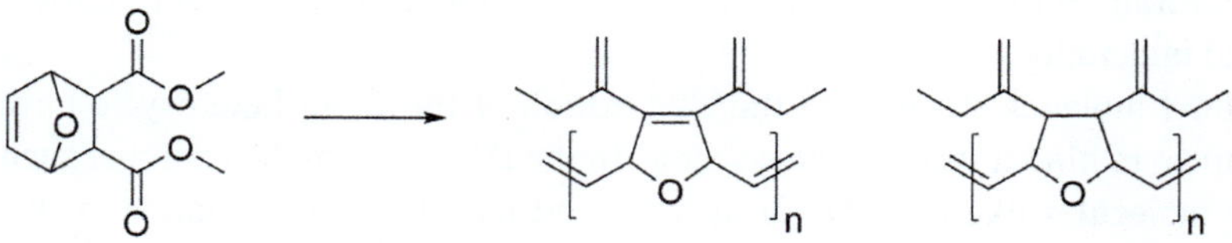

Figure 6.52 ROMP of a modified furan/maleic anhydride DA adduct and the polymer obtained from the reduction of the original adduct.

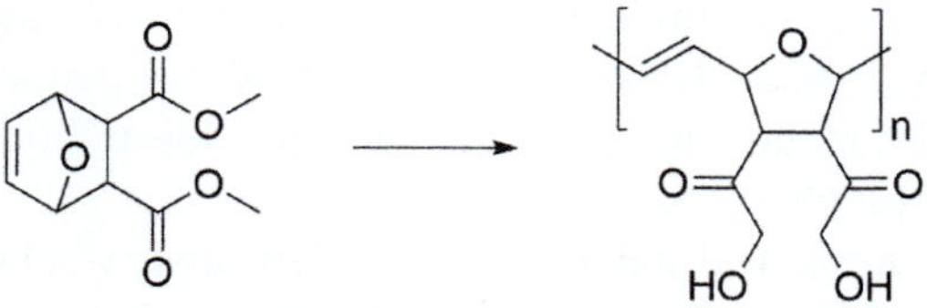

Figure 6.53 ROMP of a hydrolyzed furan/maleic anhydride DA adduct.

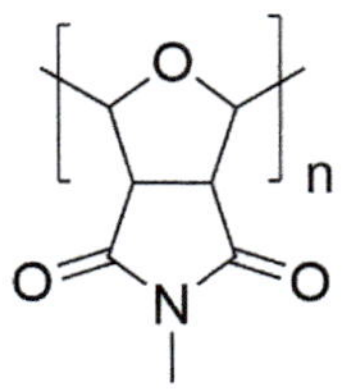

Figure 6.54 Polymer formed in the ROMP of the furan/*N*-methylmaleime DA adduct.

mode, again using $RuCl_3$ as catalyst, giving polymers with molecular weights ranging between 1 and 7×10^5 Da [73].

On the whole, these ROMP systems were studied with special emphasis on the kinetics, mechanisms and the structures and molecular weights of the expected polymers, and less attention to the properties of the materials, except for the fact that they had a good thermal stability.

6.4 A New Paradigm: Aromatics from Furans

The conversion of the furan/maleimide adducts into aromatic structures was briefly mentioned and sketched in Figure 2.16. The DA aromatization strategy typically involves two steps, that is, cycloaddition of a furanic diene with an appropriate dienophile, followed by acid catalyzed dehydration of the intermediate DA adduct, as studied by several groups, and as recently summarized in an important contribution that provides a novel and original approach to this chemical transformation [74], and later thoroughly assessed in a comprehensive review on the DA reaction [75]. Although this topic could appear to digress from the "furan polymer" content of the book, it bears in fact a major significance because it highlights the very promising perspective of producing aromatic compounds, normally based on fossil sources, from simple and cheap furan derivatives through the exploitation of their DA adducts with dienophiles. In other words, a fresh area dealing with the preparation of new important sustainable molecules and materials is being born [75].

The novel synthetic procedure [74] is summarized in Figure 6.55 applied to 2-methylfuran and maleimide, but obviously also applies to other furan derivatives and maleimides.

Carbonyl moieties attached to the C2 position of the furan heterocycle have been shown to hinder its reactivity as diene in the DA reaction, thus making such current molecules like F, HMF, furoic acid and its esters unexploitable in this context. This generally recognized notion has been recently challenged by a series of experiments in which appropriate conditions have made it possible for some of these compounds to behave as "normal" furan dienes. In particular, F and HMF were successfully reacted with various maleimides in an aqueous medium [76] as shown in Figure 6.56. This behavior was interpreted as arising from the partial hydration of the formyl group which reduced its conjugation with the ring unsaturations and thus restored their dienic character.

The same group also proved that furoic acid and some of its C5 derivatives would react with maleimides in a basic aqueous solution [77] according to the scheme in Figure 6.57. Various furoic acid esters and amides were also successfully tested.

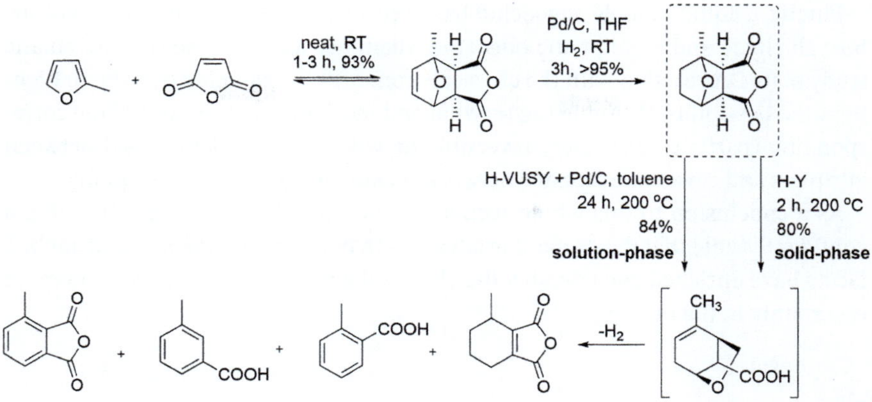

Figure 6.55 The transformation of the furan/maleic anhydride DA adduct into three aromatic compounds.

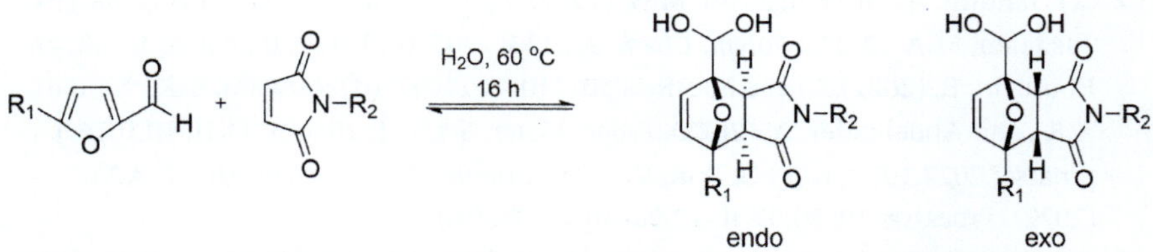

Figure 6.56 The major products in the DA reaction between furfural and some of its derivatives and various maleimides in aqueous solution. Here R^1 = H, CH_2OH, Me, CH_2OMe, and R^2 = H, Me, Et, *n*Pr, Ph.

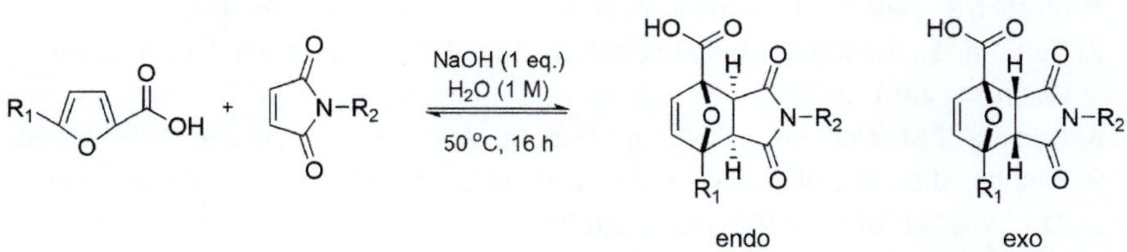

Figure 6.57 The major products of the DA reaction between furoic acid and some of its C5 derivatives and various maleimides in a basic aqueous solution. Here R^1 = H, Me, CH_2OH, COOH, and R^2 = H, Me, nPr, Ph.

In more general terms, the kinetics and thermodynamics of the furan-based DA reaction with a variety of dienophiles was critically reviewed [75] to highlight its enormous potential as a synthetic tool to prepare a vast array of aromatic compounds, which opens new avenues of obtaining these usually fossil-derived family of chemicals using cheap and abundant renewable furans, starting from first generation mono- and 2,5-disubstituted members, notably F and HMF.

Finally, a liquid, soluble oligoSchiff base, recently prepared from 2,5-furandicarboxyaldehyde and an aliphatic oligoether diamine, was the base of a systematic study of its DA reaction with the classical aromatic bismaleimide in order to highlight the possibility of tuning the thermal and mechanical properties of the corresponding (partially) thermally reversible networks [78], which ranged between vitrimers and covalent adaptable networks possessing good reprocessability.

As a conclusion to this whole section on the furan-based DA reaction, it can confidently state that the deepening advances in both its fundamental and applied facets have enriched considerably the chemical and materials sciences and more is certainly in the making.

References

1 Gandini, A. (2013). *Progr. Polym. Sci.* 38: 1–29.

2 (a) Gandini, A. and Belgacem, M.N. (1997). *Progr. Polym. Sci.* 22: 1354–1368. (b) Tasdelen, M.A. (2011). *Polym. Chem.* 2, 2133–2145. (c) Briou, B., Améduri, B., and Boutevin, B. (2021). *Chem. Soc. Rev.* 50: 11055–11097. (d) Ratwani, C.R., Kamali, A.R., and Abdelkader, A. (2022). *Progr. Mater. Sci.* 131: 101001. DOI: 10.1016/j.pmatsci.2022.101001. (e) Galkin, K.I., Sandulenko, I.V., and Polezhaev, A.V. (2022). *Processes* 10: 30. DOI: 10.3390/pr10010030.

3 Gandini, A., Coelho, D., and Silvestre, A.J.D. (2008). *Eur. Polym. J.* 44: 4026–4033.

4 Canadell, J., Fischer, H., De With, G., and Van Benthem, R.A.T.M. (2010). *J. Poym. Sci. A Polym. Chem.* 48: 3456–3467.

5 Lorenzini, R.G. and Sotzing, G.A. (2014). *J. Appl. Polym. Sci.* 131: 40179–40186.

6 Motoki, S., Nakano, T., Tokiwa, Y., et al. (2015). *Polymer* 101: 98–106.

7 Gandini, A., Coelho, D., and Silvestre, A.J.D. (2010). *J. Polym. Sci. Part A Polym. Chem.* 48: 2053–2060.

8 Lacerda, T.M., Carvalho, A.J.F., and Gandini, A. (2016). *RSC Adv.* 6: 45696–45700.

9 Banella, M.B., Giacobazzi, G., Vannini, M., et al. (2019). *Macromol. Chem. Phys.* 220: 1900247. DOI: 10.1002/macp.201900247.

10 (a) Chen, X., Dam, M.A., Ono, K., et al. (2002). *Science* 295: 1698–1702. (b) Chen, X., Wudl, F., Mal, A.K., Shen, H., and Nutt, S.R. (2003). *Macromolecules* 36: 1802–1807.

11 (a) Gandini, A., Coelho, D., and Silvestre, A.J.D. (2013). *Polym. Chem.* 4: 1364–1371. (b) Gandini, A., Coelho, D., and Silvestre, A.J.D. (2011). *Polym. Chem.* 2: 1713–1719.

12 Heo, Y. and Sodano, H.A. (2014). *Adv. Funct. Mater.* 24: 5261–5268.

13 Mc Elhanon, J.R. and Weeler, D.R. (2001). *Org. Lett.* 3: 2681–2683.

14 Szlai, M.L., Mc Grath, D.V., Weeler, D.R., Zifer, T., and Mc Elhanon, J.R. (2007). *Macromolecules* 40: 818–823.

15 Frayne, S.H., Stolz, R.M., and Northrop, B.H. (2019). *Org. Biomol. Chem.* 17: 7878–7883.

16 Laita, H., Boufi, S., and Gandini, A. (1997). *Eur. Polym. J.* 33: 1203–1211.

17 Chujo, Y., Sada, K., and Saegusa, T. (1990). *Macromolecules* 23: 2636–2641.

18 (a) Goussé, C., Gandini, A., and Hodge, P. (1998). *Macromolecules* 31: 314–321. (b) Gheneim, R., Perez-Berumen, C., and Gandini, A. (2002). *Macromolecules* 35: 7246–7253. (c) Gheneim, R. (2002). Doctorate Thesis, Grenoble National Polytechnic Institute, France. (d) Perez-Berumen, C. (2002). Doctorate Thesis, Grenoble National Polytechnic Institute, France.

19 Gandini, A., Carvalho, A.J.F., Trovatti, E., Kramer, R., and Lacerda, T.M. (2018). *Eur. J. Lipid Sci. Technol.* 120: 1700091. DOI: 10.1002/ejlt.201700091.

20 Elschner, T., Obst, F., and Heinze, T. (2018). *Macromol. Biosci.* 18: 1800258. DOI: 10.1002/mabi.201800258.

21 Juang, J., Dufresne, A., and Lin, N. (Eds.) (2019). *Nanocellulose: From Fundamentals to Advanced Materials*. Wiley-VCH, Weinheim, Germany.

22 Shao, C., Chang, H., Wang, M., Xu, F., and Yang, J. (2017). *ACS Sustainable Chem. Eng*. 5: 6167–6174.

23 Kramer, R.K., Belgacem, N.M., Carvalho, A.J.F., and Gandini, A. (2019). *Int. J. Biol. Macromol.* 141: 493–498.

24 Carneiro de Oliveira, J., Rigolet, S., Marichal, C., Roucoules, V., and Laborie, M.-P. (2020). *Carbohydr. Polym.* 250: 116966. DOI: 10.1016/j.carbpol.2020.116966.

25 García-Astrain, C., Gonzalez, K., Gurrea, T., et al. (2016). *Carbohydr. Polym.* 149: 94–101.

26 Huang, J., Fu, S., and Gan, L. (Eds.) (2019). *Lignin Chemistry and Applications*. Elsevier, Chemical Industry Press, Amsterdam, The Netherlands.

27 Buono, P., Duval, A., Averous, L., and Habibi, Y. (2017). *Polymer* 133: 78–88.

28 Thys, M., Brancart, J., Van Assche, G., Vendamme, R., and Van den Brande, N. (2021). *Macromolecules* 54: 9750–9760.

29 (a) Duval, A., Lange, H., Lawoko, M., and Crestini, C. (2015). *Green Chem.* 17: 4991–5000. (b) Zhou, W., Zhang, H., and Chen, F. (2018). *Int. J. Biol. Macromol.* 107: 790–795.

30 Miao, M. and Campanella, O.H. (Eds.) (2022). *Starch-based Materials: Science and Engineering*. New York, USA, Associated Press.

31 Nossa, T.S., Belgacem, N.M., Gandini, A., and Carvalho, A.J.F. (2015). *Polym. Intern.* 64: 1366–1372.

32 Fernandes Antunes, B., Ferreira, A.G., Capaldo Amaral, A., et al. (2022). *Polym. Intern.* 71: 1340–1346.

33 Trovatti, E., Lacerda, T.M., Carvalho, A.J.F., and Gandini, A. (2015). *Adv. Mater.* 27: 2242–2245.

34 Trovatti, E., Cunha, A.G., Carvalho, A.J.F., and Gandini, A. (2017). *Int. J. Biol. Macromol.* 95: 762–769.

35 Rinaudo, M. (2006). *Progr. Polym. Sci.* 31: 603–632.

36 Montiel-Herrera, M., Gandini, A., Goycoolea, F.M., et al. (2015). *Carbohydr. Polym.* 128: 220–227.

37 Garcia-Astrain, C., Gandini, A., Peña, C., et al. (2014). *RSC Adv.* 4: 35578–35587.

38 Gandini, A. and Lacerda, T.M. (2019). *Polymers from Plant Oils*. Scrivener Publishing, Beverly, MA, USA.

39 (a) Vilela, C., Cruciani, L., Silvestre, A.J.D., and Gandini, A. (2011). *Macromol. Rapid Commun.* 32: 1319–1323. (b) Vilela, C., Cruciani, L., Silvestre, A.J.D. and Gandini, A. (2012). *RSC Adv.* 2: 2966–2974.

40 Vilela, C., Silvestre, A.J.D., and Gandini, A. (2013). *J. Polym. Sci. Polym. Chem.* 51: 2260–2270.

41 Gandini, A., Lacerda, T.M., and Carvalho, A.J.F. (2013). *Green Chem.* 15: 1514–1519.

42 Lacerda, T.M., Carvalho, A.J.F., and Gandini, A. (2014). *RSC Adv.* 4: 26829–26837.

43 (a) Gevrek, T.N. and Sanyal, A. (2021). *Eur. Polym. J.* 153: 110514. DOI: 10.1016/j.eurpolymj.2021.110514. (b) Morozova, S.M. (2023). *Gels* 9: 102. DOI: 10.3390/gels9020102.

44 Shi, M., Wosnick, J.H., Ho, K., Keating, A., and Shoichet, M.S. (2007). *Ang. Chem. Intern. Ed.* 46: 6126–6131.

45 Fisher, S.A., Anandakumaran, P.N., Owen, S.C., and Shoichet, M.S. (2015). *Adv. Funct. Mater.* 25: 7163–7172.

46 Tam, R.Y., Fisher, S.A., Baker, A.E.G., and Shoichet, M.S. (2016). *Chem. Mater.* 28: 3762–3770.

47 Smith, L.J., Taimoory, S.M., Tam, R.Y., et al. (2018). *Biomacromolecules* 19: 926–935.

48 Gregoritza, M. and Brandl, F.P. (2015). *Eur. J. Pharm. Biopharm.* 97: 438–453.

49 Shi, M. and Shoichet, M.S. (2008). *J. Biomater. Sci. Polym. Ed.* 19: 1143–1157.

50 Yamashita, S., Fukushima, H., Mori, T., Katayama, Y., and Niidome, T. (2011). *Langmuir* 27: 14621–14626.

51 Shi, M., Ho, K., Keating, A., and Shoichet, M.S. (2009). *Adv. Funct. Mater.* 19: 1689–1696.

52 Gregoritza, M., Messmann, V., Abstiens, K., Brandl, K.F.P., and Goepferich, A.M. (2017). *Biomacromolecules* 18: 2410–2418.

53 Altinbasak, I., Sanyal, R., and Sanyal, A. (2016). *RSC Adv.* 6: 74757–74764.

54 Mai, B.T., Conteh, J.S., Gavilan, H., Di Girolamo, A., and Pellegrino, T. (2022). *ACS Appl. Mater. Interfaces* 14: 48476–48488.

55 Gevrek, T.N., Ozdeslik, R.N., Sahin, G.S., et al. (2012). *Macromol. Chem. Phys.* 213: 166–172.

56 Kalaoglu-Altan, O.I., Kirac-Aydin, A., Sumer Bolu, B., Sanyal, R., and Sanyal, A. (2017). *Bioconjug. Chem.* 28: 2420–2428.

57 Keller, K.A., Guo, L., and Finn, M.G. (2005). *Tetraherdon Lett.* 46: 1181–1184.

58 McElhanon, J.R., Zifer, T., Kline, S.R., et al. (2005). *Langmuir* 21: 3259–3266.

59 Dirlam, P.T., Strange, G.A., Orlicki, J.A., Wetzel, E.D., and Costanzo, P.J. (2010). *Langmuir* 26: 3942–3948.

60 Patej, D.J., Graham, K.R., and Reynolds, J.R. (2012). *J. Mater. Chem.* 22: 3004–3014.

61 Smulders, M.M.G. and Nitschke, J.R. (2012). *Chem. Sci.* 3: 785–788.
62 Costa, G., Gandini, A., Morinelli, S., and Valenti, B. (1993). *Heterocycles* 36: 1225–1237.
63 Park, M., Stricker, F., Campos, J.G., et al. (2023). *ACS Macro Lett.* 12: 33–39.
64 Don, J.-L. and Friščić, T. (2017). *ACS Central Sci.* 3: 13–19.
65 Cardosa-Gutierrez, M., De Bo, G., Duwez, A.-S., and Remacle, F. (2023). *Chem. Sci.* 14: 1263–1271.
66 Duan, Y.-X., Wang, Y.-X., Wang, L.-J., et al. (2017). *Macromolecules* 50: 1353–1361.
67 Hu, X., Zeng, T., Husic, C.C., and Robb, M.J. (2019). *J. Am. Chem. Soc.* 41: 15018–15023.
68 Zhang, M. and De Bo, G. (2020). *J. Am. Chem. Soc.* 42: 5029–5033.
69 Yoon, S., Choi, J.H., Sung, B.J., Bang, J., and Kim, T.A. (2022). *NPG Asia Mater.* 14: 61. DOI: 10.1038/s41427-022-00406-3.
70 Hamilton, J.G., Marquess, D.G., O'Neill, T.J., and Rooney, J.J. (1990). *J. Chem. Soc., Chem. Commun.* 1991: 119–121.
71 Feast, W.J. and Harisson, D.B. (1991). *Polymer* 32: 558–563.
72 Hillmyer, M.A., Lepetit, C., McGrath, D.V., Novac, B.M., and Grubbs, R.H. (1992). *Macromolecules* 25: 3345–3350.
73 (a) Lu, S.Y., Quayle, P., Heatly, F., et al. (1992). *Macromolecules* 25: 2692–2697. (b) Viswanathan, T. and Jethmalani, J. (1993). *J. Appl. Polym. Sci.* 48: 1289–1295.
74 Thiyagarajan, S., Genuino, H.C., van der Waal, J., et al. (2016). *Angew. Chem. Int. Ed.* 55: 1368–1371.
75 Cioc, R.C., Crockatt, M., van der Waal, J.C., and Bruijnincx, P.C.A. (2022). *Angew. Chem. Int. Ed.* 61: e202114720.
76 Cioc, R.C., Lutz, M., Pidko, E.A., et al. (2021). *Green Chem.* 23: 367–373.
77 Cioc, R.C., Smak, T.J., Crockatt, M., van der Waal, J.C., and Bruijnincx, P.C.A. (2021). *Green Chem.* 23: 5503–5510.
78 Lucherelli, M.A., Duval, A., and Averous, L. (2023). *ACS Sustainable Chem. Eng.* 11: 2334–2344.

7

Chemical and Biological Degradation

Whereas the highlighted peculiar chemical features of the furan heterocycle, notably its dienic character, have been richly exploited to prepare original materials with properties sometimes unattainable when using monomers and polymers either from fossil resources, or from non-furan renewable ones, they are also responsible for detrimental effects on the stability of the macromolecular materials we have discussed throughout this book. First, it is instructive to examine some typical lack of chemical stability of furan compounds before dealing with their polymers. We have already shown in Section 2.4 that the hydrolysis of the heterocycle only becomes a relevant problem when a strongly acidic aqueous medium is employed. Likewise, its photochemical decomposition is only operational at wavelengths shorter than those present under ordinary atmospheric or artificial illumination, unless the ring is further conjugated with C=C and/or carbonyl groups, and even then, the photochemical transformations involve those side moieties and can therefore be exploited in useful applications, as outlined in Section 2.3 and in numerous subsequent examples applied to polymers, like dimerization reactions. A more drastic situation is illustrated by the instability of such compounds as 2-furfuryl chloride, where its self-ionization produces a reactive furfuryl cation which attacks the C5 position of another molecule, giving rise to a sequence of condensations resulting in colored resins following the same pathway as that discussed at length in the acid-catalyzed polycondensation of 2-furfuryl alcohol (FA). Atmospheric oxidation reactions of furan compounds without the use of a specific catalyst have been shown to be negligible, even at moderately high temperatures, unless free radicals are present in the medium.

Some of these conditions can however become more relevant in polymers incorporating the furan ring. A typical example is the slow loss of solubility of poly(2-vinyl furan), which is caused by free-radical crosslinking. Here, the loss of the tertiary hydrogen atom shown in Figure 7.1 leaves a polymer radical stabilized by the furan diene, which thereafter will slowly couple with another one and thus

Furan Polymers and their Reactions, First Edition. Alessandro Gandini and Talita Martins Lacerda.

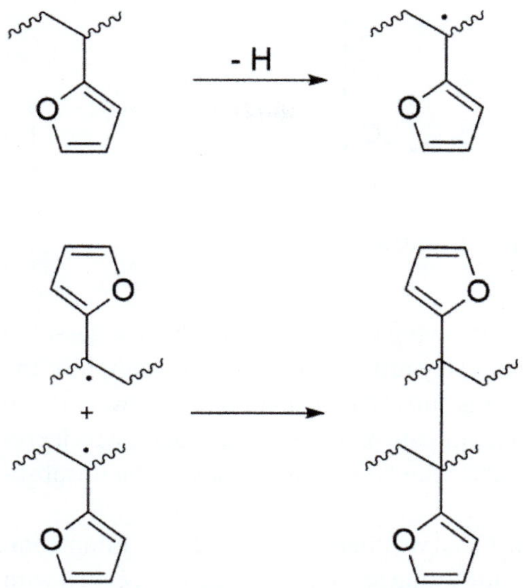

Figure 7.1 The sequence of initial events leading to the crosslinking of poly(2-vinyl furan) exposed to the atmosphere and hence to its loss of solubility.

leads to the ultimate formation of a network. Other reactions can also occur in this context, directly involving the heterocycle, as briefly outlined in Section 2.1.

This fragility is not limited to free radical hydrogen abstraction, but also applies to the ease of proton and hydrogen ion loss in basic and acidic media, respectively, because the ensuing carbanion and carbocation are stabilized by the furan diene system, as clearly argued in such polymer reactions as the acid-catalyzed polycondensation of FA (-CH_2 moieties between two furan rings) and of alkenyl furans bearing a tertiary hydrogen atom on the carbon attached to the ring C2 position. In both instances the regular polymer structure is perturbed by the formation of unwanted polyunsaturation.

The other reactive position that can induce either desired or undesired consequences is the unsubstituted C5 position of the furan ring when operating in an acidic environment, viz. the electrophilic substitution first discussed in Section 2.2. Examples include the positive effect of this reaction, which is in fact the driving force in the polymerization of FA and, conversely, the unwanted branching in the cationic polymerization of 2-vinyl and 2-isopropenyl furan.

A puzzling observation was recently reported concerning the changes in properties of an original furan-aliphatic polyester in which 3,3'-bisfuran-5,5'-dicarboxylic acid was used as monomer [1], as shown in Figure 7.2. This polymer displayed an exceptionally high oxygen barrier, but this feature was progressively reduced with air exposure of weeks at room temperature, together with loss of crystallinity, increase in Tg and decrease in water contact angle. The authors did not examine

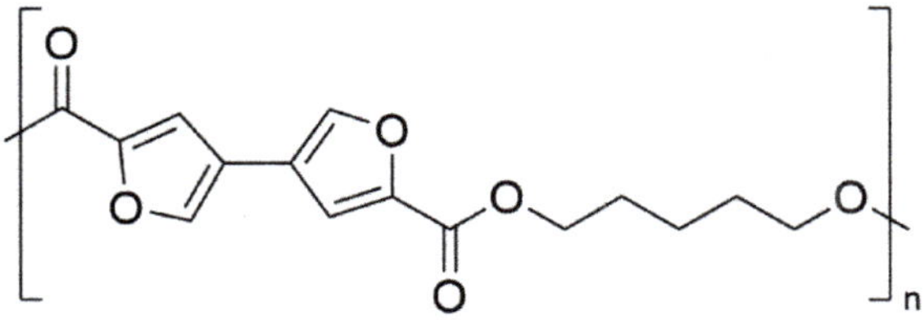

Figure 7.2 A recent furan polyester with a 3,3'-5,5'-difuran moiety.

in detail the structural changes accompanying these features, which suggest the occurrence of cross-linking, but tentatively attributed them to oxidation of the rings with formation of polar lactone groups. This is plausible and we wish to attribute this reactivity, not reported in the case of more classical polyesters incorporating 2,5-disubstituted furans (see Section 5.1), to the lability of the unsubstituted C2 positions.

The same group also studied the artificial weathering of the 2,2'-5,5'-furan homologue of the polyester in Figure 7.2 and found some degradation associated with color formation, partial cross-linking and loss of crystallinity, but no change in Tg [2], and attributed it to the higher conjugation of the bifuran dicarbonyl moiety, i.e., absorption at higher wavelengths towards the visible, compared with the 2,5-monofuran homologue. In a different context, the by-now paradigmatic furan polyester PEF (see Section 5.1.2.1) was unexpectedly reported in 2021 [3] to show sensitivity to UV irradiation in a standard weathering chamber, with coloration, chain scission, partial cross-linking and crystallinity and decrease in Tg. The authors tentatively attributed these changes to Norrish-type I and II degradation reactions, but announced further investigation on the issue. We look forward to finding the outcome of this ongoing study.

It appears that these observations on the atmospheric ageing of furan polyesters require a more thorough inspection, including the possible role of traces of catalyst remaining in the polymers, which could play a key role in some of the observed structural modifications.

Turning now to the crucial aspect related to the bio-induced degradation of polymers [4], for those bearing furan monomer units, one finds rather insufficient and often contradictory research having been carried out thus far. To the best of our knowledge, the first report on this topic described in 1997 [5] the behavior of 11 different amorphous polyesters **5a,b, 6a,b, 8a–d,** and **10a–c** synthesized by bulk polycondensations of the two difuran diesters **4a** and **4b** with two 1,4 : 3,6-dianhydrohexitols **1** and **2,** four aliphatic diols **7a–d,** and three oligo(ethylene glycol)s **9a–c** (Figure 7.3).

The authors first examined the response of the two furan monomers 4a and 4b to catabolic degradation by an activated micro-organism sludge at room temperature, and found that the evolution of carbon dioxide reached about 20% of the

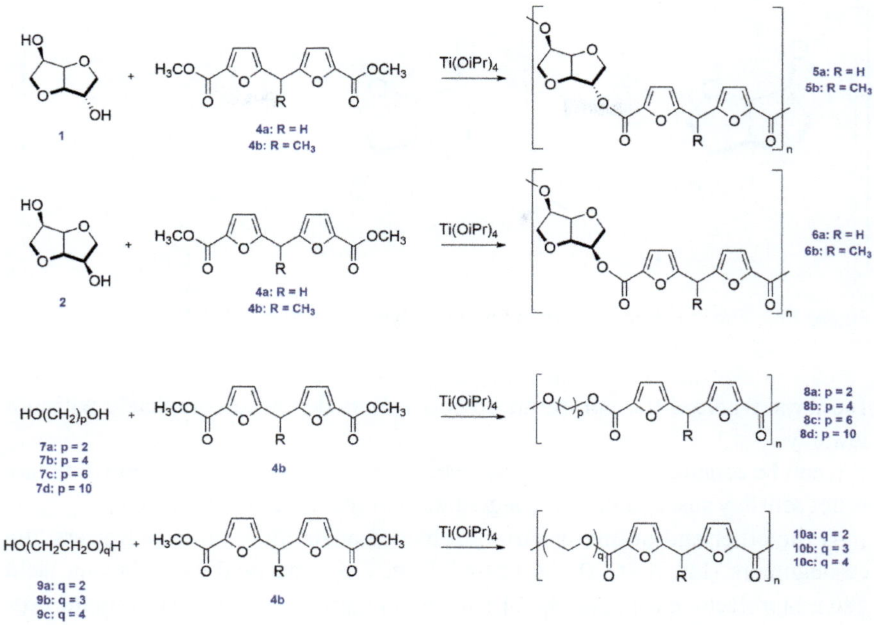

Figure 7.3 The furan polyesters first examined in terms of their biodegradability.

theoretical amount for complete carbon conversion within a month, e.g., that these compounds were degradable, albeit less rapidly than the cyclic ether monomers 1 and 2. Compost-soil degradation of the polymer films 5b and 8b indicated surface microorganism attack with modest weight loss of 10 and 2%, respectively after 10 weeks, viz. rather slow degradation. Finally, the polymers were submitted to enzymatic degradation by a lipase from porcine pancreas at 37 °C for 24 h, by measuring again the amount of evolved CO_2. The polyesters 8a–8d displayed a curious behavior in that the extent of degradation followed the order 8b > 8c > 8d > 8a, which was interpreted as the opposite role of steric hindrance and hydrophilic character, on the basis of similar trends for the polymer series 5, 6, and 10. On the whole however, the presence of furan monomers in these polyesters did not appear to show a marked aptitude to facilitate their biodegradation, as indeed confirmed in a later investigation by the same authors [6], who prepared random copolyesters of different compositions based on the structure given in Figure 7.4. Their enzymatic degradations under the action of three different lipases systematically showed a strong reduction of the CO_2 evolution with time as the percentage of furan comonomer was increased, indicating that the presence of the heterocycle resulted in a lower aptitude to undergo the degradation. The same trend was observed when the copolyesters were submitted to soil degradation, with the 100% furan homopolymer displaying a near-total resistance to erosion even after

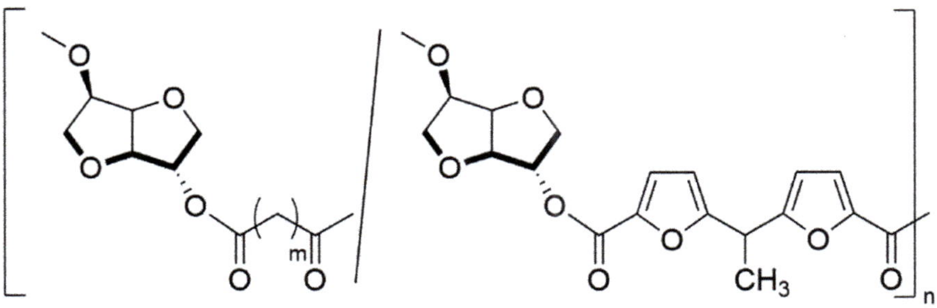

Figure 7.4 The structure of random furan copolyesters with m = 4–12.

160 days, whereas the non-furan homopolymers degraded completely within a few days.

It can be concluded from these pioneering tests that the furan-carbonyl moiety is not sensibly susceptible to biodegradation in these types of polyesters.

At the other end of this spectrum of biodegradability of furan polyesters, the co-oligomers (Mn~1000 Da) prepared from 2-hydroxymethyl-5-caboxylic acid and ε-caprolactone using lipase [7], shown in Figure 7.5, exhibited enzymatic degradations in which the presence of the furan moiety *enhanced* the susceptibility to undergo depolymerization. The same trend was observed in degradation by an aqueous medium, containing micro-organisms from a natural environment. These features are most likely due to the high concentration of -OH and -COOH end-groups in the oligomers, favoring the furan ester hydrolysis. The authors concluded that these materials were more suitable for biomedical applications than the poly(ε-caprolactone) homopolymer.

An interesting example of conflicting results is found in two publications seemingly reaching opposite evidence concerning the biodegradability of copolyesters from butylene adipate and 2,5-furandicarboxylate. In the first study [8], the enzymatic degradation was strongly reduced by the increasing percentage of furan monomer units to the point of inhibition at 75%. Two years later, a brief report on the compost soil degradation of the same copolymers [9] showed efficient and essentially similar degradation rate for 5, 10, and 20% of furan units, modestly favoring the latter composition, with the aliphatic homopolymer being somewhat more resistant. Although the two degradation tests were of a different biological nature, these results are surprising and highlight the lack of systematic investigations concerning the biodegradability of furan-bearing polymers. In a different investigation, the degradation of copolyesters based on lactic acid, FDCA, and ethylene glycol [10] was monitored in both compost soil and a phosphate buffer solution, providing evidence of the slowing role of the presence of furan monomer units in the weight loss of the samples and in the molecular weight decrease of the recovered materials.

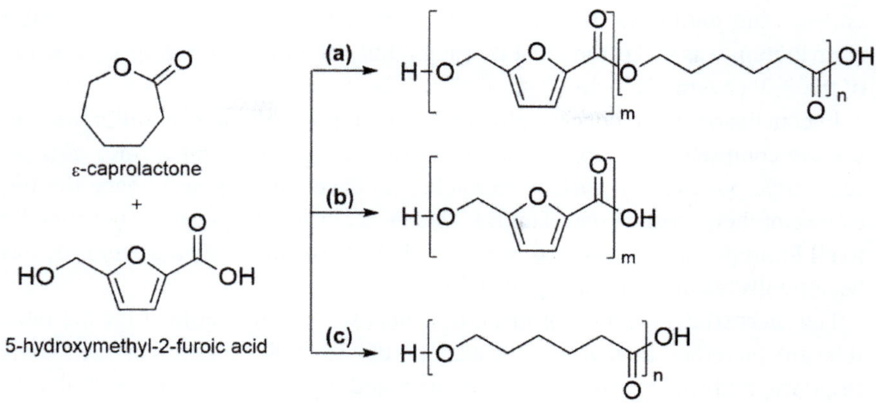

Figure 7.5 The monomers and oligoesters tested for biodegradation.

Regrettably, few studies deal specifically with the biodegradability of PEF, the most important furan polyester of industrial interest. The simple hydrolytic degradation at pH 6.9 and 37 °C resulted in a very minor weight loss of some 2% within 12 weeks [11]. The enzymatic hydrolysis of PEF samples of different molecular weights (6, 10, and 40 kDa) by cutinase 1 from *Thermobifida cellulosilytica cutinase* was monitored by measuring the amount of FDCA released after 8, 24, 48, and 72 h at 50 °C, which displayed a linear increase [12], the other major products of this degradation being monoesters and dimers. In the absence of the enzyme, PEF did not show any detectable change at these operating conditions. In a similar investigation which however used both PEF films and powders of different average size and added a second enzyme to that described above [13], highlighted the accelerating role of the polymer amorphous character and the fact that cutinases from *Humicola insolens* operating at pH 8 and 65 °C degraded the amorphous polymer entirely within 72 h. These results were reiterated in another report by the same groups with the only relevant addition of comparative data related to PEF and PET, which highlighted that the former degrades 1.7 times more rapidly than the latter under the same enzymatic conditions [14]. A less explicit information was provided by the Avantium company in 2019, giving results of compost soil degradation of PEF obtained by the Organic Waste System in Ghent, Belgium, showing the "unweathered" PEF degraded to about 95% within 450 days, whereas a "weathered" sample took 250 days to reach 90% degradation [15]. Using the same conditions, PET displayed less than 10% degradation in both situations, although details about the "weathering" process were not provided. In 2018, a study described the use of a novel enzyme, PETase double-mutant enzyme, capable of degrading crystalline PET using a pH 7.2 phosphate buffer at 30 °C all the way to the monomers [16]. The authors briefly extended this original investigation to the degradation of crystalline PEF by the same enzyme and found that its activity was even higher than with PET. This latest contribution was by far the most thorough, but still limited to the specific action of the PEF enzymatic hydrolysis.

In conclusion, the biodegradation of PEF and its copolyesters is still in need of a more comprehensive approach, considering the growing importance of these renewable polymers in various technological developments and hence the relevance of their possible recyclability. This need obviously applies more generally to all furan polymers discussed in this book, because on the whole very little has been published on their biodegradability.

The thermal degradation of furan polymers was briefly outlined for the most relevant materials in the various sections of the book in terms of their TGA response, and their general behavior suggested a good stability compared with that of aromatic and aliphatic counterparts, with the notable exceptions of the polymer modifications mentioned above because of the hydrogen-atom lability in monomer units such as those of poly(2-vinylfuran). In other words, no particular fragility was detected in the temperature ranges associated with typical processing operations and, in many instances, the presence of the heterocycle resulted in an enhanced stability. FA resins are an exception, not related to polymer fragmentation, but instead to enhanced cross-linking and tendency to graphitization at the higher temperatures, which was exploited for the manufacture of carbonaceous applications like industrial electrodes, electrically conductive films produced by 3D printing, and rigid foams (see Section 3.2).

References

1 Kainulainen, T.T., Parviainen, T.A.O., Sirvio, J.A., McGeachie, L.J.R., and Heiskanen, J.P. (2013). *ACS Macro Lett.* 12: 147–151.

2 Kainulainen, T.T., Godwa, V., Heiskanen, J.P., and Hedenqvist, M.S. (2022). *Polym. Degr. Stab.* 200: 109960. DOI: 10.1016/j.polymdegradstab.2022.109960.

3 Maaskant, E. and van Es, D.S. (2021). *ACS Macro Lett.* 10: 1616–1621.

4 (a) RameshKumar, S., Shaiju, P., O'Connnor, J.E., and Babu, P.R. (2020). *Curr. Opin. Green Sust. Chem.* 21: 75–81. (b) Okada, M. (2002). *Progr. Polym. Sci.* 27: 87–133. (c) Hatti-Kaul, R., Nilsson, L. J., Rehnberg, N., and Lundmark, S. (2020). *Trends Biotechnol.* 38: 50–67.

5 Okada, M., Tchikawa, K., and Aoi, K. (1997). *J. Polym. Sci. A: Polym. Chem.* 35: 2729–2737.

6 Okada, M., Tchikawa, K., and Aoi, K. (1999). *J. Appl. Polym. Sci.* 74: 3342–3350.

7 Todea, A., Bitcan, I., Aparaschivei, D., et al. (2019). *Polymers* 11: 1402. DOI: 10.3390/polym11091402.

8 Zhou, W., Wang, X., Yang, B., et al. (2013). *Polym. Degr. Stab.* 98: 2177–2183.

9 Jacquel, N., Saint-Loup, R., Pascault, J.-P., Rousseau, A., and Fenoullot, F. (2015). *Polymer* 59: 234–242.

10 Wu, H., Wen, B., Zhou, H., et al. (2015). *Polym. Degr. Stab.* 121: 100–104.

11 Matos, M., Sousa, A.F., Fonseca, A.C., et al. (2014). *Macromol. Chem. Phys.* 215: 2175–2184.

12 Pellis, A., Haernvall, K., Pichler, C.M., et al. (2016). *J. Biotechnol.* 35: 47–53.

13 Weinberger, S., Canadell, J., Quartinello, F., et al. (2017). *Catalysts* 7: 318. DOI: 10.3390/catal7110318.

14 Weinberger, S., Haernvall, K., Scaini, D., et al. (2017). *Green Chem.* 19: 5381–5384.

15 Gruter, G.-J. (2019). Technology and Markets Day Path to the Future. Available online at: https://www.avantium.com/wp-content/uploads/2019/06/20190606. Accessed February 3, 2023.

16 Austin, H.P., Allen, M.D., Donohue, B.S., et al. (2018). *PNAS* 115: E4350–E4357.

8

General Conclusions

The fact that the number of academic and industrial research groups working on the multifarious aspects of furan polymers has been steadily increasing in the last couple of decades is a clear indication of the relevance of the field within the equally growing interest in the general area of polymers from renewable resources. Our deliberate choice to include in this book both the less successful investigations and the promising outcome of others, was to alert the reader about the pitfalls associated with the peculiar chemical behavior of the furan heterocycle, as frequently highlighted throughout the book. This does not imply that certain ventures should be avoided, but rather that if certain tricky systems seem however worth tackling, then it should only be done with a clear knowledge of the underlying chemistry and hence only with particularly original strategies. One example of this situation is the inherent instability of the $(2,5\text{-Fu-CH}_2)_n$ polymer derived from the polycondensation of FA seems unavoidable, but given the objective evidence that this material would be extremely interesting and its precursor particularly cheap and abundant, a clever idea could arise aiming at an intelligent way to protect the regular growing macromolecule from its degeneration with an appropriate stabilizing modification carried out *in situ* during the process. This working philosophy is brilliantly sustained by the recent proof that 2-furan-carbonyl moieties, generally thought to be unreactive as the diene in the DA coupling with dienophiles, can indeed be made to respond to that condensation by returning to relatively simple and clever chemical adaptations, which in fact should have been thought of decades before.

We made a point of insisting about the need for further work when the systems discussed were sufficiently promising, but incompletely studied and will therefore not elaborate any further with a list of such systems. It is however important to provide some concluding remarks on what we consider the most rewarding present or as yet unexplored lines of research. If, on the one hand, the interest in furan-based polyesters and copolyesters is presently in full swing, the corresponding polyamides, on the other hand, lag behind, considering in particular the

Furan Polymers and their Reactions, First Edition. Alessandro Gandini and Talita Martins Lacerda.

search for KEVLAR-like materials, which were only modestly investigated more than 30 years ago. Likewise, furan-based polyethers have received very scanty attention without convincing reasons that would justify such neglect. In the light of the disastrous "plastic pandemic", the urgent need of research on the recyclability and biodegradation of furan polymers is worth reiterating here, although one can speculate that these investigations are indeed taking place confidentially in industrial laboratories. In more general terms, much remains to be done of the properties of the polymers discussed in this book, covering mechanical, electrical, electronic, morphological, and optical performances, because their applications depend strictly upon one or more of these features.

The question that this book could inspire is whether its contents show that virtually all the obvious furan polymers have already been explored, even if in a scanty approach, or whether entirely new structures are still waiting to be studied. We have some ideas and hope that the book will likewise inspire some of its readers to tackle that challenge.

Index

Furan Polymers and their Reactions, First Edition. Alessandro Gandini and Talita Martins Lacerda.

d

t

u